Progress in
Psychobiology and
Physiological Psychology

Volume 17

Progress in PSYCHOBIOLOGY AND PHYSIOLOGICAL PSYCHOLOGY

Edited by ADRIAN R. MORRISON

Department of Animal Biology
School of Veterinary Medicine
University of Pennsylvania
Philadelphia, Pennsylvania

STEVEN J. FLUHARTY

Department of Animal Biology
School of Veterinary Medicine
University of Pennsylvania
Philadelphia, Pennsylvania

Volume 17

ACADEMIC PRESS
San Diego • London • Boston
New York • Sydney • Tokyo • Toronto

This book is printed on acid-free paper. ∞

Academic Press
a division of Harcourt Brace & Company
525 B Street, Suite 1900, San Diego, California 92101-4495, USA
http://www.apnet.com

Academic Press Limited
24-28 Oval Road, London NW1 7DX, UK
http://www.hbuk.co.uk/ap/

International Standard Book Number: 0-12-542117-6

PRINTED IN THE UNITED STATES OF AMERICA
98 99 00 01 02 03 OW 9 8 7 6 5 4 3 2 1

Contents

The Locus Coeruleus–Noradrenergic System as an Integrator of Stress Responses

Rita J. Valentino, Andre L. Curtis, Michelle E. Page, Luis A. Pavovich, Sandra M. Lechner, and Elisabeth Van Bockstaele

A Model for the Control of Ingestion—20 Years Later

John D. Davis

Contributors

Numbers in parentheses indicate the pages on which the authors' contributions begin.

Andre L. Curtis, Department of Psychiatry, Allegheny University, Philadelphia, Pennsylvania 19129-1191 (91)

John D. Davis, E. W. Bourne Behavioral Research Laboratory, New York Hospital/Cornell Medical Center, White Plains, New York 10605 (127)

Sandra M. Lechner, Allelix Neuroscience, Inc., South Plainfield, New Jersey 07080 (91)

Timothy H. Moran, Department of Psychiatry and Behavioral Sciences, Johns Hopkins University School of Medicine, Baltimore, Maryland 21218 (1)

Michelle E. Page, Department of Psychiatry, University of Pennsylvania, Philadelphia, Pennsylvania 19104 (91)

Luis A. Pavcovich, Department of Psychiatry, Allegheny University, Philadelphia, Pennsylvania 19129-1191 (91)

Jay Schulkin, Department of Physiology and Biophysics, Georgetown University, Washington, DC 20007-2197 (35)

Gary J. Schwartz, Department of Psychiatry and Behavioral Sciences, Johns Hopkins University School of Medicine, Baltimore, Maryland 21218 (1)

Priyattam J. Shiromani, Veterans Administration Medical Center and Harvard Medical School, Brockton, Massachusetts 02401 (67)

Rita J. Valentino, Department of Psychiatry, Allegheny University, Philadelphia, Pennsylvania 19129-1191 (91)

Elisabeth Van Bockstaele, Department of Pathology and Anatomy, Thomas Jefferson University, Philadelphia, Pennsylvania 19107 (91)

Preface

In Volume 17 of *Progress in Psychobiology and Physiological Psychology* we return to our standard format, presenting a series of essays by workers in various areas of behavioral neuroscience eager to tell a story about a problem that has consumed them. Our previous volume, of course, focused on the career of former co-editor Alan Epstein and offered articles by various students and colleagues of Alan who wanted to remember him. The entire series, though, offers opportunities for our authors to range widely with their ideas and present them in any style they wish. Our only requirement is that each essay relate the brain to behavior and behavior to the brain, a tradition that originated with the founding editors, Jim Sprague and Eliot Stellar.

In his chapter, Priyattam (Peter) Shiromani champions the use of molecular techniques to advance the field of sleep research. He reviews the recent evidence of immediate-gene activation during sleep and wakefulness, emphasizing that, as with all techniques, results with c-*fos* must be interpreted with care. Drawing from other areas of neuroscience research, feeding behavior and circadian rhythms, he foresees that sleep researchers can look forward to unraveling sleep–wake regulation as a coordinated interaction between extracellular and intracellular/molecular events.

Rita Valentino and colleagues have written about the locus coeruleus (LC), a nucleus that has featured heavily in sleep research lore. They look at LC from a quite different perspective though. Their interest is the role of the LC–norepinephrine system in the integration of neuronal activity of diverse brain regions in response to external and internal stimuli. Although excitatory amino acids and corticotropin-releasing hormones are likely mediators of LC activation by stressors, the authors caution that the search must go on. They suggest that the LC–norepinephrine system may have a "global influence on the stress response, perhaps facilitating the integration between certain cognitive, immune, and autonomic responses." Future understanding of this sytem will likely lead to treatment of debilitating stress-related diseases, such as depression, posttraumatic stress disorder, and irritable bowel syndrome.

In a related essay, "Fear and Its Endocrine Basis," Jay Schulkin argues that a function of glucocorticoid hormones concerns the facilitation of the synthesis of corticotropin-releasing hormone in the amygdala. Both philosophical and physiological issues related to fear are addressed in this very engaging chapter. Assembling various pieces of evidence from his own work and that of others, he presents the hypothesis "that neuropeptides such as CRH code chemically the sense of fear, which is sustained by elevated cortisol."

The remaining two chapters of this series shift focus from the neural substrates of emotional behavior to that of motivation and, specifically, the control of food intake. Jack Davis reevaluates his model for the control of ingestion, which was originally published more than 20 years ago. When it was first published, the model was constructed according to the principles of control theory that were commonly applied to biological systems. The benefits of hindsight are clearly evident in Davis's present reformulation of the model. Indeed, this critical review of his own research, and that of other investigators, reveals many of the deficiencies of the original model. For instance, the earlier conception that negative feedback control of ingestion rate arose entirely from the intestines is clearly inadequate. Moreover, subsequent research has indicated the important role of conditioning in feeding behavior, and his new model attempts to incorporate learning and memory in the neural control of ingestion. Rather than being discouraged by the shortcomings of his earlier model, Davis is appropriately optimistic that the new model will stimulate additional research on the controls of eating behavior. In view of his clear conceptualization of the challenging problems that remain, we agree.

In Volume 8 of this series published in 1979, Smith and Gibbs wrote an authoritative review of postprandial satiety. The basic premise was that the neuroendocrine signals that terminate feeding and produce satiety were likely to be more amenable to experimental analysis than those that elicit ingestion. They also proposed in this essay that cholecystokinin (CCK), an intestinal hormone released by food stimuli, acted as a satiety signal. In the intervening years, it has become clear that CCK is part of the neuroendocrine system that participates in postprandial satiety. However, defining CCK's relation to other physiological cues associated with meal termination has been elusive. In their chapter, Gary Schwartz and Timothy Moran provide compelling evidence that CCK normally interacts with other gastrointestinal signals to induce satiety. The mechanism(s) of this interaction involves the identification of visceral neurons that are responsive to gastric distension, nutrients, and CCK. These neural afferents represent important sites for multiplicative interactions between, and integration of, these diverse satiety signals. Thus, their chapter has widespread implications, for it establishes that neural/hormonal synergies likely represent a more general principle employed in the control of a wide range of homeostatic behaviors.

Steven J. Fluharty

Contents of Recent Volumes

Integrative Gastrointestinal Actions of the Brain-gut Peptide Cholecystokinin in Satiety

Gary J. Schwartz and Timothy H. Moran

Department of Psychiatry and Behavioral Sciences
Johns Hopkins University School of Medicine
Baltimore, Maryland 21205

I. Introduction

In 1973, Gibbs, Young, and Smith (1973a) demonstrated that peripheral administration of either a synthetic of a partially purified form of the peptide cholecystokinin (CCK) reduced food but not water intake in rats. The pattern of behaviors by which CCK reduced food intake was consistent with the production of satiety. CCK reduced meal size and meal duration and resulted in an earlier appearance of a behavioral sequence of satiety similar to that seen following ingestion of a normal size meal (Antin *et al.*, 1975). From these data, Gibbs, Young, and Smith (1973) hypothesized that CCK was a satiety hormone that was released from the gastrointestinal tract by the presence of ingested food and that it served as a negative feedback signal involved in the termination of food intake. Since that time, the feeding inhibitory actions of CCK have been demonstrated in a variety of species, incuding nonhuman primates and humans, and with the availability of potent and specific CCK antagonists, many aspects of the original hypothesis have been confirmed. The overall significance of CCK mechanisms in the control of food intake is reinforced by recent findings associating the lack of functional CCK_A receptors with the development of obesity and diabetes in a strain of rats (Funakoshi et al., 1995) and with the incidence of a familial obesity in humans (Miller et al., 1995).

The past decade has seen significant progress in our understanding of the peptidergic control of feeding behavior, and much of this progress stems from studies evaluating the actions of CCK. This chapter reviews recent advances in our understanding of the pharmacology and molecular biology of CCK and its receptors and in the physiological effects of CCK in the context of its actions as a satiety peptide. In presenting these data, we will also consider (1) how CCK's multiple, integrative actions at gastrointestinal and peripheral neural sites may contribute to the negative feedback control of food intake and (2) how evaluation of this peptide's actions may elucidate some new general principles of brain-gut communication critical to the control of food intake.

II. CCK as a Brain-gut Peptide

CCK is found in both the brain and the gastrointestinal (GI) tract and plays a number of roles centrally and peripherally as a signaling molecule. In the brain,

0363–0951/98 $25.00

CCK is found in high but variable concentrations and has met a variety of the criteria for designation as a neurotransmitter. CCK is also widely distributed throughout the GI tract, but it is most concentrated in the duodenum and the jejunum. Intestinal pools of CCK are present in duodenal endocrine cells (Buchan *et al.*, 1978) and in enteric nerves (Schulzenberg *et al.*, 1980). Intestinal CCK can be shown to be released into the bloodstream by the intraluminal presence of the digestive products of protein and fats, where it has a number of physiological, gastrointestinal functions, including the stimulation of pancreatic secretion, gallbladder contraction, intestial motility, and the inhibition of gastric emptying. Some of these functions of intestinally derived CCK are mediated by endocrine actions of the peptide, whereas others appear to depend on paracrine and/or neurocrine modes of action.

III. Site of Action of CCK in Satiety

A. CCK Receptor Subtypes

Two pharmacologically and molecularly distinct CCK receptor subtypes have been identified and characterized. The possibility of heterogeneity in CCK receptors was first suggested in the original homogenate radioligand studies on the pancreas and brain (Innis and Snyder, 1980), where the relative affinity of some CCK analogs for the binding sites differed between these two tissues. Nonsulfated CCK, pentagastrin, and CCK-4 bound with high affinity to the brain. In contrast, these CCK analogs had relatively low affinity for CCK receptors found in pancreatic tissue homogenates.

Autoradiographic studies demonstrated that CCK binding sites in the brain are both widely and variably distributed (Zarbin *et al.*, 1983). A combination of autoradiographic studies and analyses of pharmacological specificity of binding demonstrated that brain CCK binding sites were heterogeneous (Moran *et al.*, 1986). Two distinct profiles of CCK binding were found at different brain sites. One profile was that which had been demonstrated in binding to cortical homogenates—both sulfated CCK-8 and nonsulfated CCK analogs interacted with this receptor population with a relatively high affinity. In contrast, binding to several discrete brain regions demonstrated the binding profile that had been found in pancreatic homogenates. Binding at these sites was relatively insensitive to the presence of unsulfated CCK, gastrin, or various CCK fragments. From this work, the present nomenclature for the two CCK receptors was proposed (Moran *et al.*, 1986). The designation CCK_A was proposed for the receptor subtype initially localized in pancreatic homogenates that preferentially bound sulfated CCK over other analogs. The CCK_B designation was proposed for the receptor subtype that was originally characterized in homogenates from the cerebral cortex.

The availability of potent and specific CCK antagonists has greatly facilitated the pharmacological characterization of CCK receptors at various sites. Hill and Woodruff (1990), using specific CCK antagonists as radioligands, confirmed the existence of both CCK_A and CCK_B receptor subtypes in the brain and extended the identification of the number of brain sites containing CCK_A receptors. The overall distribution of these two CCK receptor subtypes in the rat is now well documented (Table 1).

Recently, CCK_A and CCK_B receptor proteins have been isolated and cloned, and the amino acid sequences of these receptors have been identified (Wank *et al.*, 1992a; Kopin *et al.*, 1992). Both CCK_A and CCK_B receptors consist of seven transmembrane-spanning protein domains and are members of the G-protein coupled superfamily of receptor molecules. The CCK_A receptors identified in rat brain in autoradiographic studies have now been shown by *in situ* hybridization probes to be the same receptor as rat pancreatic CCK_A receptors (Honda *et al.*, 1993). Furthermore, CCK_B receptors and gastrin receptors represent a single gene product (Lee *et al.*, 1992). The distribution of the two CCK receptor subtypes appears to be species specific. For example, there is no evidence of CCK_A receptor gene expression in human pancreas. All human pancreatic CCK receptors appear to be CCK_B receptors (Wank, Pisegna, and de Weerth, 1994).

B. Pharmacological Characterization of the Actions of Exogenous CCK in Satiety

The development of potent and specific CCK_A and CCK_B receptor agonists and antagonists (Table 2) has greatly facilitated our understanding of CCK's actions. Two pharmacological strategies have been employed to evaluate the receptor subtypes mediating CCK's actions in satiety. The first has evaluated the ability of CCK_A and CCK_B agonists and antagonists to mimic or block the satiety actions of exogenously administered CCK. Early studies comparing the ability of sulfated and nonsulfated CCK to inhibit food intake demonstrated that only the sulfated form had satiety actions (Gibbs, Young, & Smith, 1973b; Lorenz *et al.*, 1979). Similarly, in studies

TABLE 1
LOCATIONS OF CCK RECEPTORS

CCK_A	CCK_B
Gastrointestinal tract	
Pancreas	Pancreas
Gall bladder	Stomach
Pylorus	
Peripheral nervous system	
Afferent vagus	Afferent vagus
Brain	
Nucleus tractus solitarius	Wide distribution
Area postrema	
Interpeduncular nucleus	
Median raphe	
Dorsal medial hypothalamus	
Nucleus accumbens	

TABLE 2
RELATIVE AFFINITY OF CCK AGONISTS AND ANTAGONISTS FOR CCK_A AND CCK_B RECEPTORS

IC_{50} (nmol/l)	CCK_A	CCK_B
CCK agonists		
CCK-8 0.1	0.3	
CCK-8US	600	2.6
CCK-4 5330	2.6	
A-71623	3.7	4400
A-63387	6300	0.7
CCK antagonists		
Proglumide	6 300 000	11 000 000
Lorglumide (CR-1409)	7.1	500
Devazepide (L-364 718 MK-329)	0.2	31
L-365 260	240	5.2
CI-988 4300	1.7	
Cam 1481	2.8	260

Note. Data compiled from Friedinger (1992), Hill *et al.* (1992), and Asin *et al.* (1992b).

using specific CCK agonists, Asin and colleagues (1992a, b) have shown that a CCK_A-receptor-selective CCK-tetrapeptide, A-71623, caused a dose-dependent suppression of food intake, whereas peripheral administration of a dose range of a selective, high affinity CCK_B-specific agonist failed to produce satiety. Together, these results suggest that CCK's interactions with the CCK_A receptor subtype mediate the satiety elicited by peripheral exogenous administration of CCK. Consistent with these findings, CCK_A antagonists dose-dependently and competitively block the satiety effects of exogenously administered CCK in a number of settings, including sham feeding (Melville, Smith, and Gibbs, 1992), during the dark phase of ad lib, real feeding of solids and liquid nutrients (Reidelberger, Varga, and Solomon, 1991; Reidelberger & O'Rourke, 1989), and during liquid sucrose or glucose consumption in a daytime test (Dourish *et al.*, 1989a; Moran *et al.*, 1992). As shown in Figure 1, increasing doses of the CCK_A antagonist devazepide (L-364,718) in rats dose-dependently blocked the inhibition of food intake produced by intraperitoneal administration of 4 micrograms per kilogram body weight (μg/kg) CCK, but administration of a dose range of the CCK_B antagonist L-365,260 was ineffective (Moran *et al.*, 1992). These findings of CCK_A mediation of exogenous CCK-induced satiety using devazepide as a CCK_A antagonist have been extended to other mammalian species, including mice (Silver *et al.*, 1989) and cats (Bado *et al.*, 1991).

IV. High and Low CCK_A Affinity States and Satiety

Early CCK receptor binding studies of the pancreas indicated that CCK was binding to two distinct sites: one with a high affinity for CCK and one with a low affinity. Wank, Pisegna, and de Weerth (1994) have demonstrated that these two

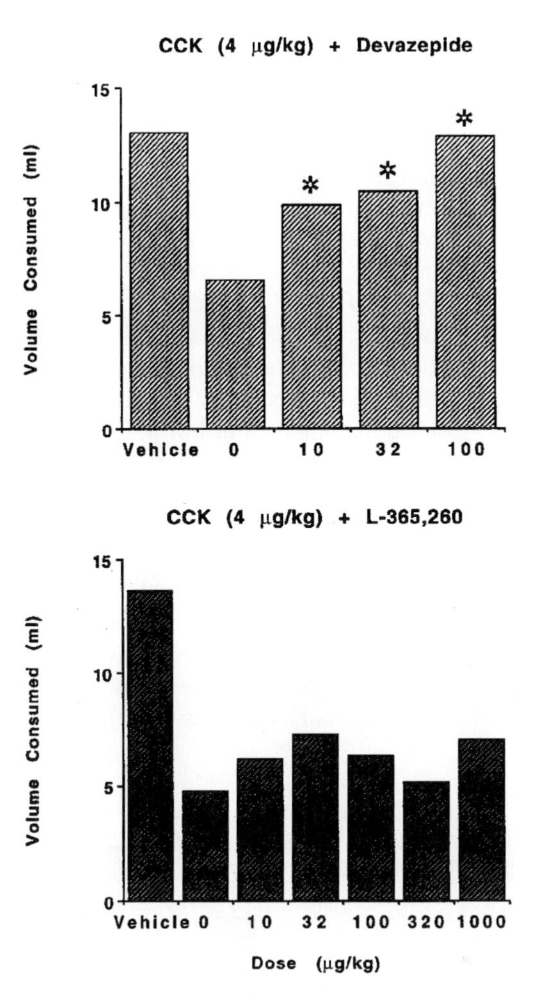

FIGURE 1 Effect of a dose range of the CCK_A receptor antagonist devazepide (upper panel) or the CCK_B receptor antagonist L-365,260 (lower panel) on 4 µg/kg CCK-induced suppression of liquid 12.5% glucose consumption during a 60 minute test in rats. Asterisks indicate statistically significant differences ($P < 0.05$) from the vehicle condition.

sites represent two different conformational states of the same CCK_A receptor protein rather than distinct receptors. Activation of these two receptor states in functional assays appears to correspond to distinct stimulatory and inhibitory actions on pancreatic amylase release. Thus, occupation of the high affinity site by low concentrations of CCK (less than 100 picomolar concentration = 1×10^{-12} molar [pM]) is correlated with stimulation of amylase release, whereas occupation of a low affinity site by concentrations of more than 100 pM is correlated with suppression of amylase secretion (Sankaran *et al.*, 1980: Sankaran *et al.*,

1982). A heptapeptide analog of CCK (CCK-JMV-180), a synthetic analog of the C-terminal heptapeptide of CCK, has the unique pharmacological profile of being a functional agonist at high affinity sites and a functional CCK_A antagonist at low affinity sites (Fulcrand *et al.*, 1988; Galas *et al.*, 1988). That is, it has the same efficacy as CCK for stimulating pancreatic amylase secretion, but it does not inhibit amylase release at supramaximal concentrations. However, these supramaximal concentrations of CCK-JMV-180 will provide a dose related blockade of the ability of higher concentrations of CCK to inhibit amylase secretion.

This distinct profile of CCK-JMV-180 makes it a valuable tool in assessing whether various actions of CCK are mediated through interactions with low or high affinity CCK receptor sites. We (Weatherford *et al.*, 1993) and others (Asin *et al.*, 1992a) addressed the issue of whether the feeding inhibitory actions of exogenous CCK were mediated through high or low affinity CCK_A receptors. The logic of the experiments was to ask whether CCK-JMV-180 was a functional CCK agonist or antagonist in a feeding paradigm. In rats, intraperitoneal CCK JMV-180, when administered alone, failed to act as a CCK agonist as it had no effect on food intake. However, CCK-JMV-180 was a functional antagonist in that it dose-dependently reversed the suppression of food intake produced by exogenously administered CCK (Figure 2). These data demonstrated that the satiety actions of exogenously administered CCK are mediated through CCK's interaction with low affinity CCK_A receptor states. Thus the examination of the pharmacological profile of the satiety actions of exogenous CCK led to the conclusion that

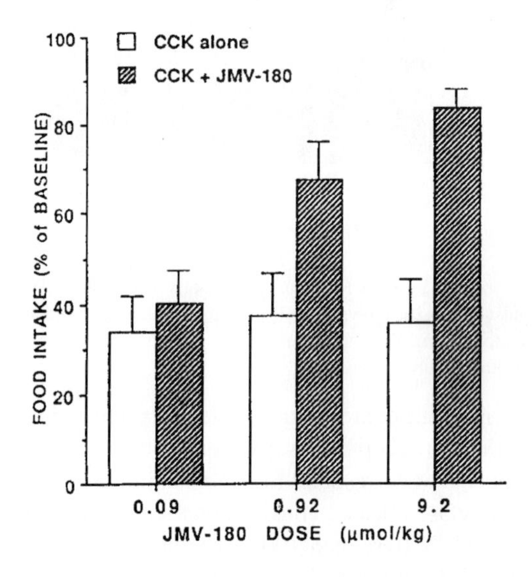

FIGURE 2 Effect of a dose range of the CCK analogue CCK-JMV-180 on 8.5 nanomoles per kilogram (nmol/kg) CCK-8 induced suppression of powdered chow intake in overnight-food-deprived rats (Weatherford *et al.*, 1993).

CCK's effects on food intake were mediated through its interactions with low affinity CCK_A receptors.

V. Role for Endogenous CCK in Satiety

A second strategy used in evaluating the actions of CCK in satiety has been to assess CCK antagonists to determine a physiological action of CCK in the control of food intake and to identify the receptor subtype mediating this satiety action of endogenous CCK. Early work by Shillabeer and Davison (1984) using the relatively nonspecific and low potency CCK antagonist proglumide demonstrated that the feeding inhibitory actions of a consumed preload could be reduced in the presence of this antagonist. Rats that ate a large amount of food in the preload would consume very small amounts of food in a subsequent food intake test situation. Following proglumide administration, rats' food intake in the subsequent test situation was significantly increased. This result was interpreted to suggest that the antagonist blocked the satiety produced by endogenous CCK released in response to the food consumed in the preload. This early result with proglumide was not replicated by other investigators (Schneider, Gibbs, and Smith, 1986).

The availability of more potent and selective CCK antagonists permitted a more rigorous evaluation of whether the feeding inhibitory actions of exogenously administered CCK mimicked the action of the endogenously released peptide, and, if so, which CCK receptor subtypes mediated the satiety produced by an endogenous release of CCK. Both L-364-718 and L-365,260 are benzodiazepine derivatives whose duration of action is on the order of several hours (Lotti and Chang, 1989; Lotti *et al.*, 1987) compared to the relatively short half-life of CCK in circulation, estimated to be on the order of several minutes (Liddle *et al.* 1994). Thus, the application of these compounds produces a long-lasting antagonism of CCK_A and CCK_B receptors, respectively. The majority of data from studies with CCK antagonists reveal that the CCK_A rather than the CCK_B receptor type mediates this satiety action. However, in the first study comparing the potency of a CCK_A and a CCK_B receptor antagonist to stimulate food intake, Dourish, Rycroft, and Iverson (1989b) found that peripheral administration of either the CCK_A antagonist devazepide or the CCK_B antagonist L-365,260 promoted feeding in partially sated rats, but that L-365,260 was significantly more potent than devazepide, suggesting that the satiety actions of endogenous CCK were mediated by its interactions with CCK_B receptors. In contrast, subsequent experiments have demonstrated that devazepide, but not L-365,260, increases liquid and solid food intake in the rat in a variety of experimental settings (Reidelberger, Varga, and Solomon, 1991; Weller, Smith, and Gibbs, 1990; Moran *et al.*, 1992). For example, in rats, increasing doses of devazepide, but not L-365,260, 30 minutes prior to the presentation of a glucose solution produced significant increases in glucose intake (Figure 3). Similarly, CCK_A antagonists have been demonstrated to block the ability of intestinal nutrient infusions to inhibit sham feeding (Green-

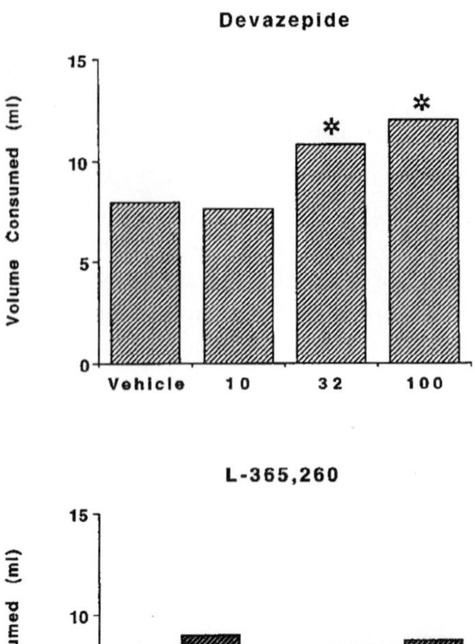

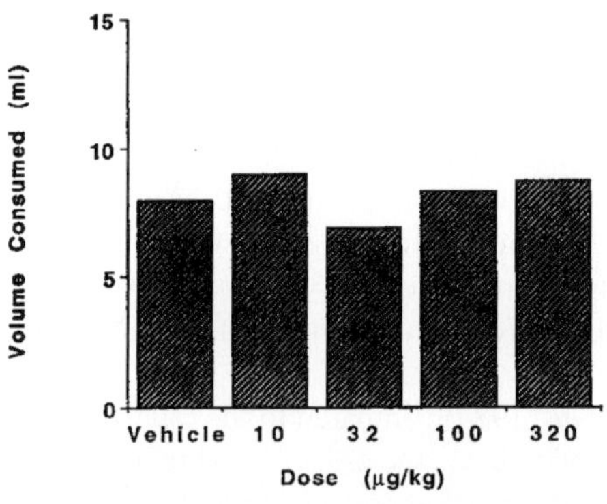

FIGURE 3 Effect of a dose range of the CCK_A receptor antagonist devazepide (upper panel) or the CCK_B receptor antagonist L-365,260 (lower panel) on liquid 12.5% glucose consumption during a 60 minute test in rats. Asterisks indicate statistically significant differences ($P < 0.05$) from the vehicle condition.

berg *et al.*, 1989; Yox, Brenner, and Ritter, 1992). Devazepide also produced dose related increases in intake in mice in a variety of testing paradigms ranging from nonfasted mice consuming chow or sucrose to prefed mice consuming a milk diet (Silver *et al.*, 1989; Weatherford, Chiruzzo, and Laughton, 1992).

In rhesus monkeys (Moran *et al.*, 1993a), intragastric administration of a dose range of devazepide following a 20-hour food deprivation resulted in dose-related increases in intake of 1 gram pellets over a 4-hour test period (Figure 4). This effect was maximal at a dose of 100 µg/kg, and represented a 40% increase over baseline levels of food intake. As in the rat, this effect was specific to administration of the CCK_A antagonist. A similar intragastric dose range of the CCK_B antagonist L-365,260 had no effect on the monkeys' food intake (see Figure 4). Analysis of the pattern of food intake during the 4-hour test situation revealed that the overall in-

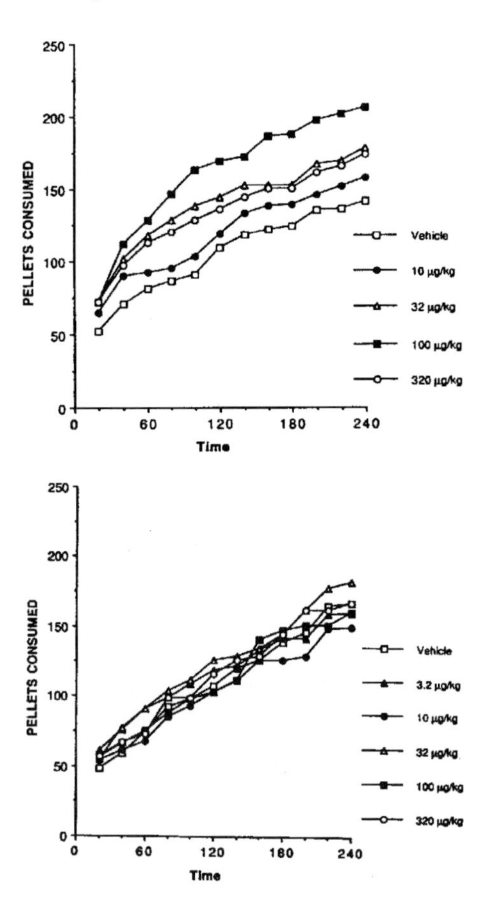

FIGURE 4 Effect of a dose range of the CCK_A receptor antagonist devazepide (upper panel) or the CCK_B receptor antagonist L-365,260 (lower panel) on pellet intake over a 4-hour test period in 20-hour food-deprived rhesus macaques (Moran *et al.*, 1993a).

crease in intake was secondary to an increase in first meal size, first meal duration, and a subsequent decrease in satiety ratio (the time interval between the end of the first meal and the beginning of the second meal divided by the size of the first meal). Together, these data have led to the confident conclusion that endogenous, meal-elicited CCK acts at CCK_A receptor sites as a physiological satiety agent.

The findings that selective blockade of CCK_A receptors promotes feeding in the absence of exogenous CCK as well as eliminating the ability of exogenous CCK to produce satiety (1) demonstrate a physiological role for endogenously released CCK in the production of satiety and (2) validate the theoretical perspective that has guided this work—that exogenous CCK activates a pathway normally stimulated by the endogenous peptide. In light of the differential receptor distribution of CCK_A and CCK_B receptor types, and the relative inability of exogenously administered CCK to penetrate the blood–brain barrier (Passaro *et al.*, 1982), these ideas have directed the search for target sites for the satiety action of CCK toward CCK_A receptor populations in the abdominal cavity. The remaining sections of this chapter will assess the potential modes of action of CCK in satiety, identify CCK's actions at various gastrointestinal CCK_A receptors, and evaluate results of experiments designed to reveal how CCK's actions at these sites may mediate or contribute to CCK satiety.

A. Mode of Action of CCK in Satiety

In their original hypothesis that CCK is a satiety peptide, Gibbs, Young, and Smith (1973a) proposed that CCK acted as a *hormone* to produce satiety. Evidence has accumulated from a number of types of studies to suggest that this aspect of the hypothesis was incorrect. Rather than affecting food intake through an endocrine mechanism, CCK's satiety actions appear to be through paracrine or neurocrine actions of the peptide. This conclusion was first suggested from an evaluation of whether postprandial plasma levels of CCK were sufficient for satiety to occur. Reidelberger and Solomon (1986) demonstrated that intravenous doses of CCK required to inhibit food intake in the rat were three to five times larger than those required for maximal pancreatic enzyme secretion, an action of CCK that was regarded as physiological and mediated by an endocrine mechanism. From these results they argued that plasma levels sufficient to inhibit food intake were unlikely to occur in response to a meal, and thus the inhibition of food intake by exogenous CCK was not mimicking an endocrine action of the endogenous peptide. This conclusion is consistent with the findings that the satiety actions of exogenous CCK are mediated through low affinity CCK receptors while the CCK interacts with the high affinity receptor state to stimulate pancreatic amylase release. Thus, levels of CCK that maximally stimulated amylase secretion would not be sufficient to activate low affinity-satiety producing sites. More recently, Reidelberger and colleagues (1994) have shown that immunoneutralization of circulating CCK blocks meal-stimulated pancreatic secretion by not satiety. Therefore, CCK must be able to derive its satiety effects through a nonendocrine mechanism.

This conclusion is further strengthened by findings of Brenner, Yox, and Ritter (1993). They were able to dissociate the ability of an intestinally infused nutrient to suppress sham feeding from its ability to increase plasma levels of circulating CCK in two ways. Specifically, they showed that (1) intraintestinal nutrients that suppressed food intake failed to elicit increases in plasma CCK levels and (2) intraintestinal nutrient infusions that did elicit increases in plasma CCK levels did not necessarily suppress food intake. Thus, nutrient-evoked local rather than plasma CCK appears to mediate CCK satiety. Together, these results support a paracrine or neurocrine, rather than an endocrine, action of CCK in satiety.

B. CCK-Induced Inhibition of Gastric Emptying

The ability of CCK to inhibit gastric emptying has been demonstrated in a variety of species, including rats, cats, nonhuman primates, and humans. For example, as illustrated in Figure 5, intravenous infusion of CCK induces a prompt and significant inhibition of the ongoing gastric emptying of physiological saline from the primate's stomach. Gastric emptying remains inhibited for the duration of the CCK infusion, and, at the termination of the infusion, gastric emptying rapidly returns to baseline levels.

Moran and McHugh (1982) had originally suggested that exogenous CCK's ability to produce satiety was completely attributable to, and secondary to, its ability to inhibit gastric emptying and promote gastric distention signals that terminated feeding. This idea was based on findings that exogenously administered CCK elicits a dose-related inhibition of gastric emptying, and doses of CCK which by themselves were unable to inhibit food intake resulted in a powerful inhibition when combined with a gastric saline preload. Although several studies in rats, humans, and other primates have demonstrated synergistic actions between CCK and gastric loads in the inhibition of food intake (Moran and McHugh, 1982; Muurhainen *et al.*, 1991; Schwartz *et al.*, 1991a), it has become clear that aspects of CCK's inhibitory actions on food intake are independent of CCK's gastric inhibitory actions. For example, CCK inhibits sham feeding where ingested food drains from the stomach (Gibbs, Young, and Smith, 1973b), and there has not been a good temporal correlation between CCK's gastric inhibitory and feeding inhibitory actions (Conover, Collins, and Weingarten, 1988). Thus, our initial strong hypothesis cannot be supported.

Experiments aimed at directly quantifying the role of gastric inhibitory influences in CCK satiety have suggested that this action depends on both gastric and nongastric actions of CCK (Moran and McHugh, 1988). CCK inhibits feeding and gastric emptying, and the magnitude of these effects are positively correlated. However, at low doses of CCK, there is an initial inhibition of feeding that occurs

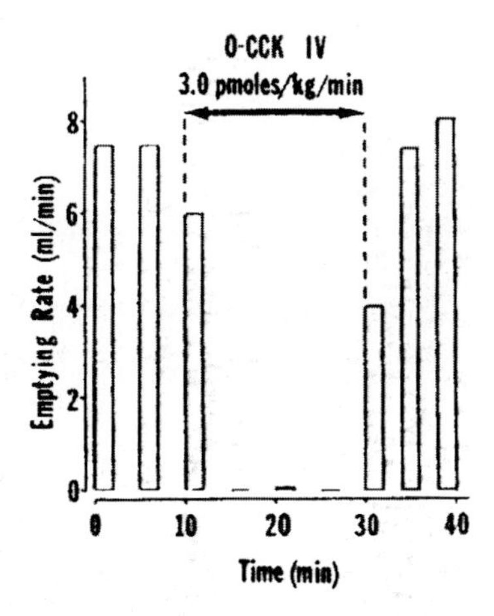

FIGURE 5 Temporal characteristics of inhibition and recovery of gastric emptying of saline that results from an intravenous infusion of CCK-8 in the rhesus macaque (Moran *et al.*, 1982).

without a concomitant inhibition of gastric emptying. Above this initial inhibitory action, a sizable portion of the satiety effect of CCK appears to be due to the gastric inhibitory actions of CCK. Thus, CCK satiety appears to depend on nongastric as well as gastric mechanisms.

Like the satiety effects of CCK, the gastric inhibitory actions of exogenously administered CCK can be shown to mimic a physiological action of endogenously released CCK. Furthermore, the pharmacological specificity of endogenous and exogenous CCK's inhibition of gastric emptying indicate these effects are, like changes in feeding, also mediated by CCK's interactions with CCK_A receptors. As illustrated in Figure 6, increasing doses of the CCK_A antagonist devazepide elicit dose-related increases in the gastric emptying of an intragastrically infused fat test meal. In the absence of the CCK_A antagonist devazepide, approximately 60% of a 0.5 kilocalories per milliliter solution (kcal/ml) fat test meal (Intralipid) remained in the primate stomach at the end of a 10-min emptying period. Beginning at a dose of 32 μg/kg, devazepide accelerates gastric empting, and at a dose of 320 μg/kg, fat empties from the monkey's stomach as if it were physiological saline. The ability of devazepide to reverse the slow emptying of nutrient infusions depends on the macronutrient composition of the test meal. For example, devazepide only partially accelerates the gastric emptying of a protein (4.5% peptone) test meal and its effects on carbohydrate meals only appear to occur when the concentration of the carbohydrate is at 300 milliosmoles/liter (mOsm).

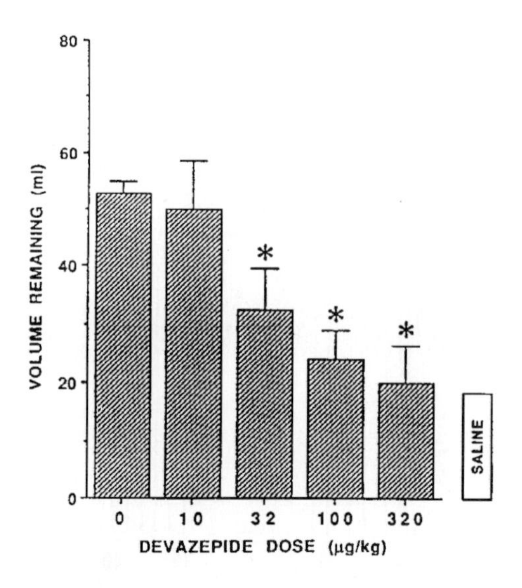

FIGURE 6 Effect of a dose range of the CCK_A receptor antagonist devazepide on the mean volume of a 100-ml Intralipid test meal (0.5 kilocalories per milliliter [kcal/ml]) remaining in the stomach at the end of a 10-minute emptying period in rhesus monkeys. Asterisk indicates a significant difference from the vehicle control (0 microgram per kilogram [μg/kg] dose) (Moran *et al.*, 1993b).

Experiments in rats using devazepide to accelerate the slowed emptying of nutrients have also underscored a role for endogenous, nutrient-elicited CCK, acting as CCK_A receptors, in the inhibition of protein, fat, and carbohydrate gastric emptying. For example, Forster and associates (1990) have shown that devazepide significantly accelerates the slow emptying of 4.5% peptone test meals. Devazepide partially accelerates gastric emptying of 10% Intralipid test meals (Hölzer *et al.*, 1994) and completely blocks the slowed gastric emptying of 300 mOsm maltose in rats (Raybould and Hölzer, 1992). Although such results clearly indicate a role for endogenous CCK in the control of liquid gastric emptying in the rat, the full extent of endogenous CCK's gastric inhibitory actions at various nutrient concentrations or with mixed nutrient meals has yet to be determined.

Because we and others have demonstrated that the satiety effects of exogenous CCK were mediated through CCK's interactions with CCK_A low affinity receptor states (see the preceding discussion), we wanted to determine whether CCK-induced inhibition of gastric emptying in rats was mediated through its interactions with the same receptor state. Again, if CCK-JMV-180 mimicked the actions of CCK on gastric emptying, this would demonstrate gastric emptying was mediated by CCK's interactions with high affinity CCK receptor states. If CCK-JMV-180 blocked the actions of CCK, this would indicate mediation at a low affinity site. As presented in Figure 7, administration of CCK-JMV-180 alone had no effect on the gastric emptying of a saline test meal. In contrast, administration of CCK inhibited gastric emptying, and administration of CCK-JMV-180 in combination with CCK resulted in a dose-dependent blockade of CCK-induced suppression of gastric emptying (Moran *et al.*, 1994). These data demonstrate that for CCK's inhibition of gastric emptying, CCK-JMV-180 acts as an antagonist of exogenous CCK.

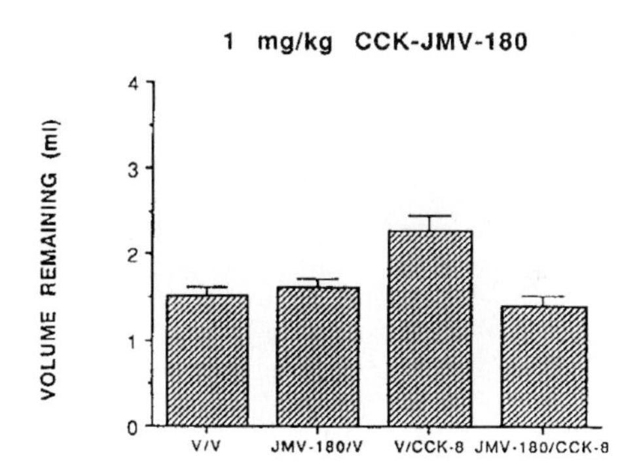

FIGURE 7 Effect of the CCK analogue CCK-JMV-180 (1 milligram per kilogram [mg/kg]) on CCK-8 induced suppression of the gastric emptying of a 5 ml bolus of physiological saline. Volume remaining in the stomach 5 min after infusion of the 5 ml gastric saline load is expressed on the y-axis. CCK (2 μg/kg) alone suppressed gastric emptying, and CCK-JMV-180 significantly reversed this suppression (Moran *et al.*, 1994). (V = Vehicle)

Thus, exogenous CCK's abilities to inhibit feeding and gastric emptying are both mediated through its interaction with low affinity CCK_A receptors. These results are consistent with a role for CCK's gastric inhibitory actions in CCK satiety.

VI. Candidate CCK Receptor Populations

Two candidate populations have been identified as potential mediators for the satiety and gastric inhibitory actions of CCK: pyloric and vagal CCK receptors. Our efforts to evaluate the potential roles of these two populations have produced a number of insights into how peptides may contribute to satiety.

A. Pyloric CCK Receptors

A discretely localized population of CCK receptors in the area of the distal pyloric sphincter has been demonstrated (Figure 8). Binding in this region corresponded to the area of thickening of the gastric circular muscle located in the most distal portion of the pyloric sphincter (Smith, G. T. *et al.*, 1984). Pharmacological characterization of I^{125} CCK binding to this site is consistent with the presence of CCK_A but not CCK_B receptors (Moran *et al.*, 1991b).

In functional studies, CCK has been shown to contract the pylorus in both *in vivo* and *in vitro* models. *In vivo* Murphy, Smith, and Gibbs (1987a) have demonstrated that exogenous administration of CCK increases in rat pyloric contraction. CCK also elicits dose-related increases in isotonic tension in an extrinsically den-

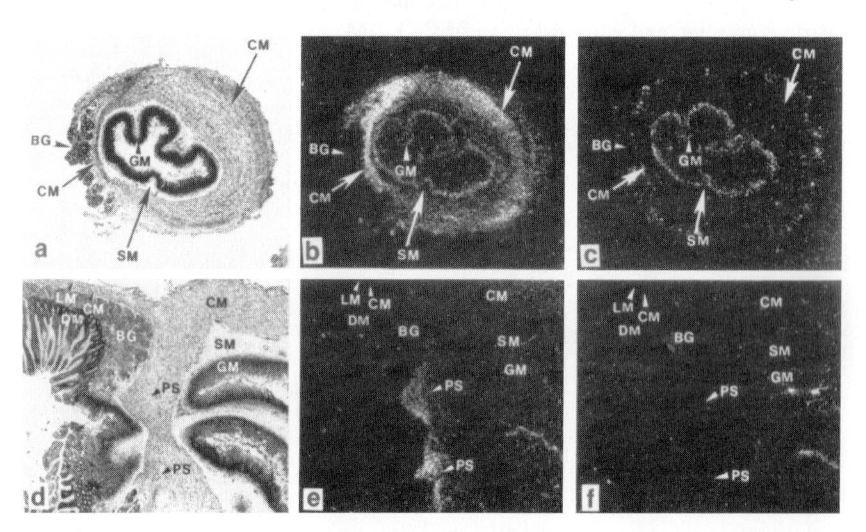

FIGURE 8 Distribution of CCK receptor sites in rat pyloric sphincter (PS). (a) toluidine blue stained cross-section; (b) autoradiographs showing total I-125 CCK-33 binding in (a) as white grains; (c) nonspecific bindings in (a); (d) toluidine blue stained longitudinal section; (e) total CCK binding in (d); (f) nonspecific binding in (d). CM = circular muscle; LM = longitudinal muscle; SM, GM, and DM = mucosal layers.

ervated, isolated pyloric sphincter preparation (Margolis, Moran, and McHugh, 1989; Murphy, Smith, and Gibbs, 1987a). In this preparation, CCK-induced pyloric contraction is also resistant to the addition of atropine and tetrodotoxin, indicating that CCK's actions do not depend on cholinergic neurotransmission, or on neural elements intrinsic to the pyloric region (Murphy, Smith, and Gibbs, 1987a). Murphy and Smith (1987b) have demonstrated that in an isolated muscle cell preparation from this region of the stomach CCK induces shortening in the individual muscle fibers, consistent with a direct action of CCK at receptors on muscle fibers.

In the isolated pyloric sphincter preparation, the pharmacological characterization of the pyloric contractile response to CCK is consistent with a role for this response in the mediation of CCK-induced satiety. Specifically, CCK-induced pyloric contraction is blocked by CCK_A antagonists (Murphy, Schneider, and Smith, 1988), and administration of CCK-JMV-180 alone had no effect on baseline tension but did block CCK-induced pyloric contraction (Moran *et al.*, 1994). (See Figure 9.) Thus, CCK's actions on the pyloric sphincter appear to be mediated through its interaction with low affinity CCK_A receptors, as are CCK's inhibitions of gastric emptying and food intake.

Removal of pyloric CCK receptors by pylorectomy results in an attenuation of only some of CCK's inhibition of gastric emptying. Following pylorectomy, CCK's ability to inhibit the emptying of a saline test meal was unaffected. However, the ability of CCK to affect the rate of emptying of a glucose test meal from the rat's stomach was attenuated (Moran, Crosby, and McHugh, 1991a). We interpreted these results as indicating that CCK affects gastric emptying through multiple sites and mechanisms. One site of action involved in the enhanced inhibition of emptying of a glucose test meal is the pylorus. However, CCK also inhibits emptying by acting at other sites and mechanisms.

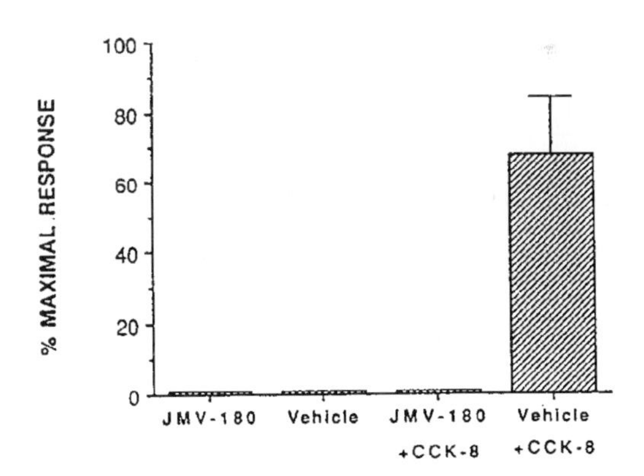

FIGURE 9 Effect of 3.2 micromolar (μM) CCK-JMV-180 on 3.2 nanomolar (nM) CCK-induced increase in pyloric contraction in *in vitro* rat pylorus (Moran *et al.*, 1994).

Pylorectomy significantly altered the ability of exogenously administered CCK to inhibit food intake, and the effects of pylorectomy on CCK satiety depend on the time of testing following the pylorectomy procedure (Moran *et al.*, 1988). As shown in Figure 10, if assessed within a few weeks of the pylorectomy, CCK-induced satiety is significantly attenuated. Specifically, the dose-reponse relationship between CCK dose and suppression of food intake is truncated following pylorectomy. The satiety produced by low doses of CCK is intact, but higher doses (4 to 8 μg/kg) are unable to produce levels of inhibition greater than those produced by the lower doses. Alternatively, if rats are tested 6 weeks to 2 months following pylorectomy, CCK dose dependently inhibits feeding across the entire dose range, as it does in intact rats (Moran *et al.*, 1988; Smith *et al.*, 1988). This renewed ability of CCK to inhibit food intake at this later time coincides with the reappearance of CCK receptors at the new gastroduodenal junction. These new CCK binding sites appear to mediate a contractile response at this time and potentially have formed a new functional pyloric sphincter.

Pyloric contraction helps determine the distribution of nutrient chyme between gastric and duodenal compartments, and in doing so inhibits gastric emptying, promotes gastric distention, and alters the pattern of duodenal exposure to nutrients. This constellation of events may in turn (1) stimulate gastroduodenal vagal chemo- and mechanosensitive afferents (see the following discussion), eliciting negative feedback signals critical for satiety and initiating gastrointestinal reflexes that further determine the pattern of nutrient delivery and (2) determine the duodenal release of endogenous CCK, which may further alter pyloric motility, inhibit gastric emptying, and suppress food intake.

Although these results demonstrate a role for pyloric CCK receptors in some of the gastric and feeding inhibitory actions of CCK, it is important to stress that py-

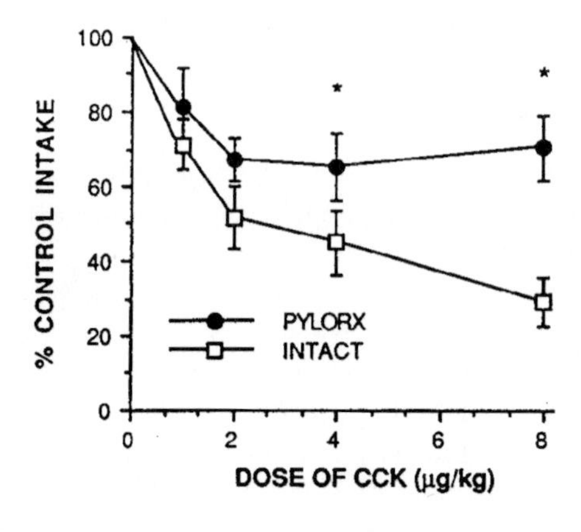

FIGURE 10 Suppression of glucose intake pre- and postpylorectomy by 1, 2, 4, and 8 μg/kg CCK in rats. Values are means ±SE (standard error). Asterisk indicates significant differences from preoperative intake (Moran *et al.*, 1988).

lorectomy attenuates, but does not eliminate, CCK-induced satiety and only affects some aspects of CCK's ability to inhibit gastric emptying. These data demonstrate that CCK must be able to produce satiety and inhibit gastric emptying by interacting with other, nonpyloric, CCK_A receptor populations. In fact, recent work of Zittel and colleagues (1995) has shown that exogenous CCK still produces satiety in gastrectomized rats. Thus, CCK has the ability to produce satiety through other CCK receptor populations.

B. Vagal CCK Receptors

The remaining major peripheral gastrointestinal site of CCK receptors identified by autoradiographic binding studies that may mediate CCK satiety is the subdiaphragmatic vagus nerve. This nerve provides the major neuroanatomical linkage between gut sites that come into contact with the nutrient products of digestion and the central nervous system substrates that mediate the control of food intake.

Zarbin and associates (1981) originally demonstrated the presence and axonal flow of CCK binding sites in the cervical vagus in rats. Subsequent experiments have revealed that transport of these CCK binding sites occurs in all of the subdiaphragmatic vagal branches (Moran *et al.*, 1987) and that transport occurs in afferent rather than efferent fibers (Moran *et al.*, 1990). The original pharmacological characterization of CCK binding in the vagus concluded that all of the binding sites under transport were CCK_A receptors (Moran *et al.*, 1987), but more recent data using both CCK agonists and antagonists have suggested that both CCK_A and CCK_B receptors are present in the vagus (Mercer and Lawrence, 1992; Lin and Miller, 1992; Corp *et al.*, 1993).

A role for the vagus in CCK satiety has been demonstrated. Although the necessity of specific subdiaphragmatic vagal branches in the mediation of CCK satiety remains unclear (Smith *et al.*, 1981; LeSauter, Goldberg, and Geary, 1988), complete subdiaphragmatic vagotomy essentially abolishes the feeding inhibitory actions of exogenous CCK (Smith *et al.*, 1981; Lorenz and Goldman, 1982; Morley *et al.*, 1982; Garlicki *et al.*, 1990).[1] Work from a number of laboratories has focused attention on the afferent vagal pathway as providing information critical to CCK satiety. Both surgical transection and chemical lesion approaches have been used to evaluate the role of this pathway. The anatomical course of the vagus nerve as it enters the rat brainstem is similar to that taken by spinal nerves; the mixed sensory and motor components in the whole vagus nerve bifurcate to form two anatomically and functionally distinct roots, one afferent and one efferent. This fortuitous anatomical arrangement has been exploited in studies designed to reveal the differential contribution of distinct vagal sensory and motor components to CCK satiety. In an influential study, Smith, Jerome, and Norgren (1985) reported that selective transection of afferent, but not efferent, vagal rootlets

[1]Reidelberger (1992) has reported that blockade of CCK_A receptors using the CCK_A receptor antagonist devazepide still elicits increases in food intake in complete subdiaphragmatic vagotomized rats, suggesting that endogenous CCK may mediate satiety through nonvagal CCK_A receptor mechanisms.

blocked the ability of a single dose of exogenous CCK to elicit satiety leading to the conclusion that the afferent vagal transection was the critical lesion in the ability of total subdiaphragmatic vagotomy to block the satiety actions of CCK.

In contrast to results from Smith, Jerome, and Norgren (1985), we have recently found that both selective total subdiaphragmatic vagal afferent and efferent transection attenuated CCK satiety, and the two procedures appeared to affect different parts of the CCK dose response curve (Moran *et al.*, 1997). As shown in Figure 11, selective transection of vagal afferent rootlets eliminated the satiety produced by low but not high doses of CCK. This result of deafferentation is similar to that produced by application of the neurotoxin capsaicin. Intraperitoneal or perivagal administration of capsaicin eliminates the satiety produced by low doses of CCK (Ritter and Ladenheim, 1985; South and Ritter, 1988), while the effects of higher doses is attenuated. Because higher doses of CCK were still able to suppress feeding after capsaicin treatment, it was suggested that CCK satiety was

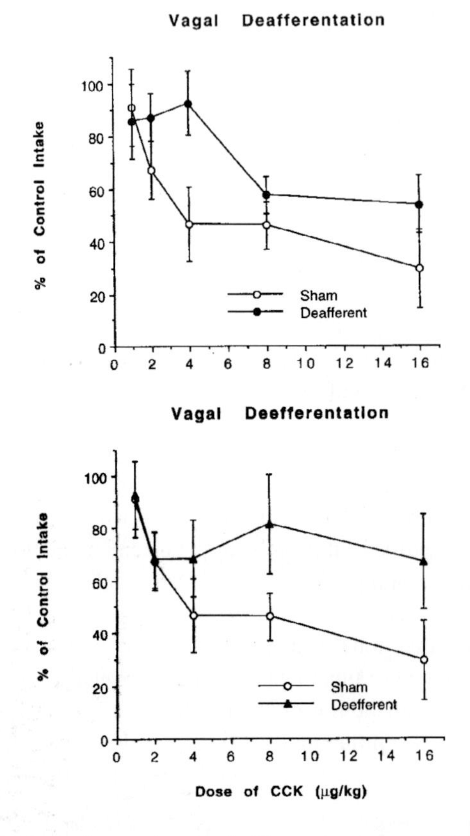

FIGURE 11 Effect of selective vagal rootlet deafferentation (upper panel) or selective vagal rootlet deefferentation (lower panel) on CCK-induced suppression of glucose intake in rats.

mediated by both capsaicin-sensitive and capsaicin-insensitive afferent fibers. The current results argue against this interpretation. Because surgical vagal afferent rootlet transection, which would completely block the contribution of both capsaicin-sensitive and noncapsaicin-sensitive afferents, produces the same pattern of results as capsaicin treatment, it appears that noncapsaicin-sensitive vagal afferents do not contribute to CCK satiety. In contrast to these results with deafferentation, vagal deefferentation eliminated the effects of high doses of CCK without affecting the suppression produced by low doses (Figure 11). This pattern of results is akin to the findings previously described for CCK satiety following pylorectomy (Table 3). We interpret this result as indicating that local gastrointestinal actions of CCK dependent on vagal efferent innervation may contribute to the suppression of food intake produced by high doses of CCK.

C. CCK-Induced Vagal Activation

Although studies using vagotomies have suggested a significant role for afferent vagal signals in CCK satiety, this approach has relied on inferences made following elimination of potentially critical vagal afferent feedback rather than direct observations of such signals. Consequently, our laboratory has adopted a strategy aimed at elucidating the nature of CCK-induced vagal afferent signals. Specifically, we have performed experiments designed to identify and characterize the meal-related neurophysiological response properties of single vagal afferent fibers. Following an observation originally made by Davison and Clarke (1988), we found that vagal afferent fibers with gastric receptive fields also increase their activity in response to close celiac arterial administration of CCK (Schwartz, McHugh, and Moran, 1991).[2] (See Figure 12.) These fibers have been characerized as slowly adapting mechanosensitive afferents. Their discharge rate is increased on load administration; increased activity is maintained for the duration that the load is held in place, and following load withdrawal there is a prototypical off-loading response—a decrease in activity below baseline levels. The excitatory neurophysiological responses to close celiac artery CCK infusions and gastric load volume are dose-related; increasing doses of CCK ranging from 1 to 1,000 picomoles (pmol) and increasing load volumes of CCK ranging from 1 to 8 ml elicit monotonic increases in firing rate in these single vagal afferent fibers (Schwartz and Moran, 1996). (See Figure 13.)

[2]Unlike our findings and those of Davison and Clarke (1988), Grundy, Bagaev, and Hillsley (1995) have reported both increases and decreases in response rate of gastric vagal load-sensitive afferents with decreases in firing rate correlated with drops in intragastric pressure. They have suggested that exogenous CCK-induced reductions in gastric pressure cause reductions in the firing rate of these gastric load-sensitive afferents. This suggestion is consistent with their contention that the mechanical state of the gastric muscular environment completely determines the firing rate of these "in series" mechanoreceptive neurons, which are sensitive to active contraction and passive distention. However, we have demonstrated that changes in global intragastric pressure and antral gastric wall tension may be dissociated from increases in vagal afferent firing rate elicited by CCK and the brain-gut peptide neuromedin C (Schwartz *et al.*, 1997). These data preclude the simple interpretation that peptide induced global gastric pressure changes uniformly elicit changes in gastric vagal afferent discharge rate.

TABLE 3
MODULATION OF CCK SATIETY

Manipulation	CCK Satiety At		Source
	Low CCK Doses	High CCK Doses	
Perivagal capsaicin	Eliminated	Attenuated	South and Ritter, 1988
Systemic capsaicin	Eliminated	Attenuated	Ritter and Ladenheim, 1985
Total subdiaphragmatic vagotomy	Eliminated	Eliminated	Smith *et al.*, 1981
Selective vagal deafferentation	Eliminated	Attenuated	Moran *et al.*, 1997
Selective vagal deefferentation	Unaffected	Eliminated	Moran *et al.*, 1997
Pylorectomy	Unaffected	Eliminated	Moran *et al.*, 1988

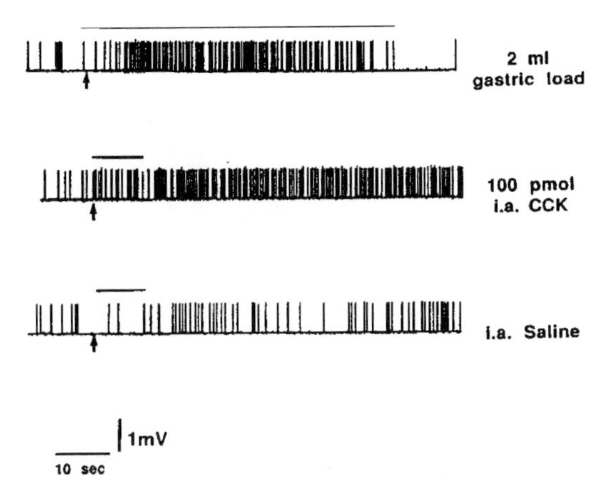

FIGURE 12 Neurophysiological recordings from a representative single vagal afferent fiber in response to a 2 ml gastric saline load (top), 100 pmol (picomoles = 1×10^{-12} moles) close celiac arterial (intraarterial [i.a.]) infusions, and saline vehicle i.a. infusions. Arrows indicate stimulus onset, and solid horizontal bars above traces indicate duration of stimulus delivery (Schwartz *et al.*, 1991b).

CCK-induced activation of these vagal afferents is mediated through interaction with CCK_A receptors (Schwartz, McHugh, and Moran, 1994). Administration of the CCK_A antagonist, devazepide, dose-dependently inhibits the response to CCK without affecting the response to intragastric loads. The interaction of the antagonist with these receptors mediating the vagal activation by CCK is competitive, in that the effect of the antagonist at higher doses of CCK produces a parallel rightward shift in the dose-response curve relating CCK dose to vagal afferent firing rate (Figure 14). In contrast, administration of the CCK_B antagonist L-365,260 across a similar dose range has no effect on the responses to either local arterial administration of CCK or to 2 ml intragastric loads.

Having demonstrated that the vagal afferent response to CCK was mediated by its interaction with CCK_A, not CCK_B receptors, we then used the CCK analog CCK-JMV-180 to determine whether these vagal afferent responses to CCK are mediated through its interaction with high or low affinity CCK_A receptor states. Because we had demonstrated that both the feeding and gastric inhibitory responses to CCK were mediated through its actions at low affinity sites, these experiments would assess whether the vagal afferent response to CCK could be playing a role in these inhibitions. As shown in Figure 15, administration of CCK-JMV-180 alone had no effect on vagal afferent activity. However, CCK-JMV-180 completely blocked the vagal afferent response to local administration of CCK. Thus, in these neurophysiological experiments, CCK-JMV-180 was a functional antagonist (Schwartz, McHugh, and Moran, 1994). Because this compound acts as an antagonist at low affinity CCK receptors, these results demonstrate that CCK elicits gut vagal afferent activity by interacting with low affinity CCK_A receptor states, a result consistent with a potential role for this vagal afferent action in the feeding and gastric inhibitory actions of CCK.

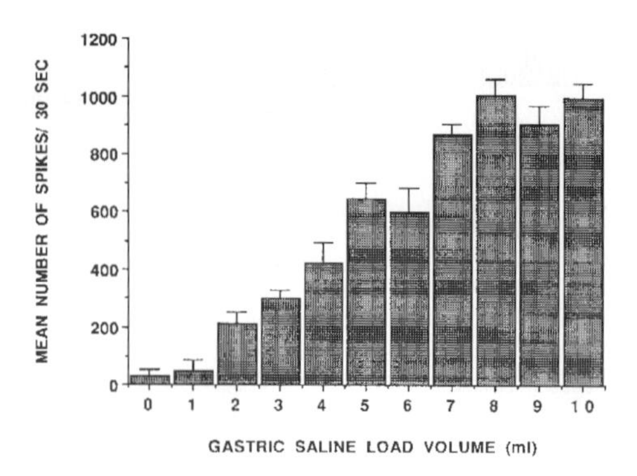

FIGURE 13 Monotonic increasing relationship between gastric load volume and spike rate in vagal mechanosensitive vagal afferents ($N = 22$) with gastric receptive fields (Schwartz and Moran, 1996).

Because endogenously released CCK inhibits gastric emptying, and thereby promotes gastric distention, it seems likely that during normal feeding, both endogenous CCK and gastric loads are present simultaneously, each potentially stimulating these vagal afferent fibers and contributing to the negative feedback control of food intake. Therefore, we examined the nature of the interaction between CCK and gastric loads in stimulating these vagal afferents. As shown in Figure 13, the vagal afferent response to gastric loads in these fibers is graded, and depending on the particular fiber, reaches a maximum at load volumes between 6 and 10 mls. Subthreshold levels of load and CCK can combine to produce vagal activation. One-ml intragastric loads that by themselves have no effect on vagal afferent activity, combined with 10 pmol of CCK, which alone has no effect, together stimulate vagal afferent activity (Figure 16). A combination of 1 ml intragastric load and 32 pmol of CCK produces a greater level of vagal activation than the combination of 1 ml intragastric load and 10 pmol of CCK. Furthermore, suprathreshold load volumes and doses of CCK also combine to produce greater

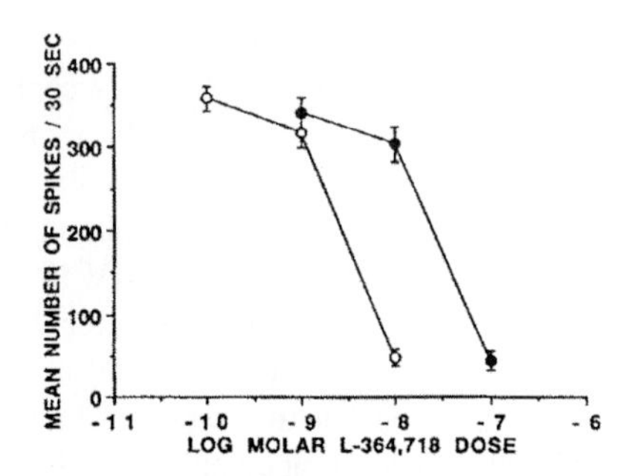

FIGURE 14 Mean number (±SE) of gastric vagal afferent neurophysiological spikes ($N = 8$ fibers) elicited by close celiac arterial infusion of 100 pmol CCK (open circles) and 1,000 pmol CCK (filled circles) in the presence of a dose range of the CCKA receptor antagonist, devazepide (Schwartz *et al.*, 1993).

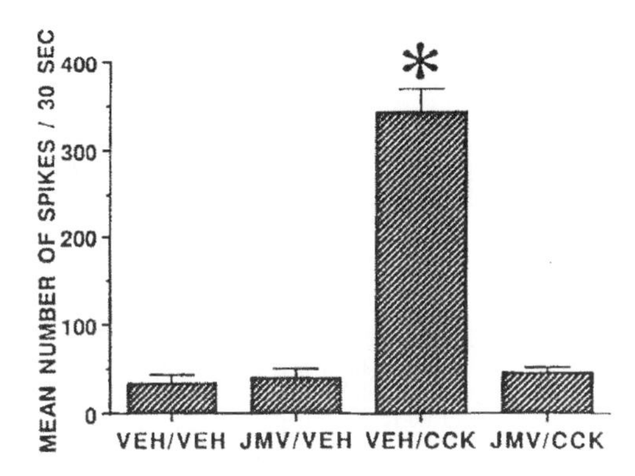

FIGURE 15 Mean number (±SE) of gastric vagal afferent spikes (N = 8 fibers) elicited by close celiac arterial infusions of saline vehicle plus antagonist vehicle (VEH/VEH), saline vehicle plus 100 nmol CCK-JMV-180 (JMV/VEH), 100 pmol CCK plus 100 nmol CCK-JMV-180 (JMV/CCK), and 100 pmol CCK plus antagonist vehicle (VEH/CCK). Asterisk indicates significant difference from (VEH/VEH) condition (Schwartz *et al.*, 1994).

levels of vagal activation than that produced by either stimulus when presented alone. At suprathreshold load volumes and doses of CCK, the combination appears to be additive in that the degree of vagal activation produced by the two in combination is roughly equivalent to the sum of the degree of activation produced by each stimulus administered individually. Thus, the vagal afferent dose-response to gastric loads is increasingly shifted to the left by pairing these loads with increasing doses of intra-arterial CCK (Figure 17). Taken together, these findings demonstrate that these vagal afferents are capable of responding to and integrating two simultaneous consequences of ingestion: the presence of CCK and the accumulation of a gastric load (Schwartz, McHugh, and Moran, 1993).

The preceding demonstrations of this synergistic interaction in gastric vagal mechanosensitive afferents suggests that these fibers may serve as a peripheral neural substrate for the behavioral integration of gastric load and CCK-induced satiety effects. Gastric loads that alone are ineffective in producing satiety in rats and nonhuman primates, or reports of fullness in humans, become effective when paired with doses of CCK that alone are ineffective in producing satiety (Moran and McHugh 1982; Schwartz *et al.*, 1991; Muurhainen *et al.*, 1991). In addition, at higher doses of CCK analogs that alone produce satiety, the suppression of food intake is enhanced in the presence of a gastric load (Schwartz, McHugh, and Moran, 1991).

Not only does CCK mimic the response to gastric load in these fibers, but it also sensitizes the fibers to subsequent gastric loads. As illustrated in Figure 18, the response to a gastric load after CCK administration and after the neurophysiological response to CCK has dissipated is significantly greater than it had been before the administration of CCK. Although we have not studied the duration of

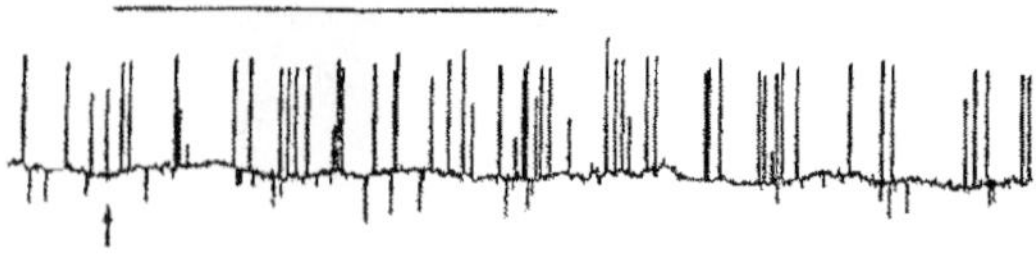

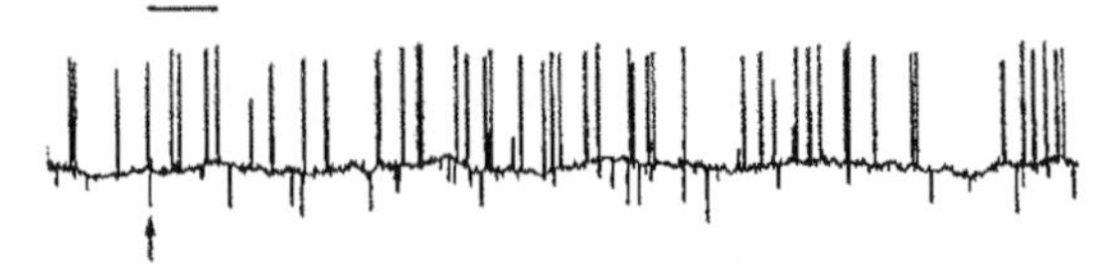

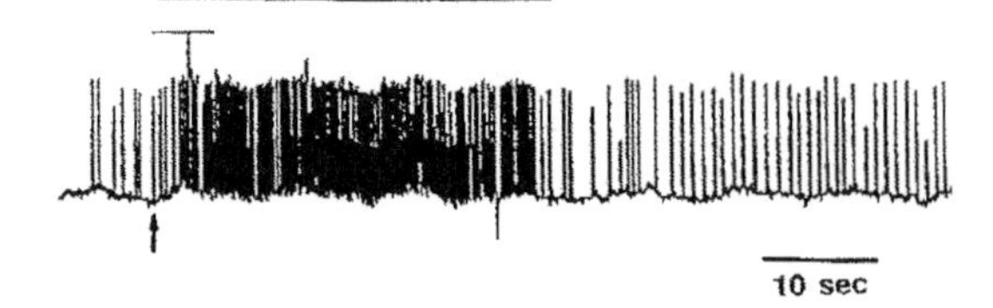

FIGURE 16 Neurophysiological recordings from a representative gastric vagal afferent fiber in response to a 1 ml-gastric saline load (A), a close celiac arterial infusion of 10 pmol CCK-8 (B), or combined administration of gastric load and close arterial CCK-8 administration (C). Arrows indicate onset of stimulus, and solid bars above traces indicate duration of stimulus delivery (Schwartz and Moran,

this sensitization systematically, individual experiments have demonstrated that it can last as long as 30 min. Thus, CCK not only mimics the excitatory effects of gastric loads in these vagal afferent fibers but also amplifies the vagal afferent response to subsequent gastric loads (Schwartz *et al.*, 1991). The mechanism underlying this sensitization is unknown.

It is not clear whether CCK excites these gastric vagal afferents by interaction with membrane bound vagal CCK receptors and/or indirectly following CCK stimulation of other peripheral CCK receptor sites. As discussed previously, CCK binding at pyloric sites stimulates pyloric contraction, and such pyloric activity can cause waves of contraction to occur in gastric sites supplied by these gastric afferents (Scheurer *et al.*, 1989). Because these gastric vagal afferents respond to both passive distention provided by gastric loads and active contracton of the gastric muscle wall, CCK-induced pyloric contraction could cause gastric contraction and indirectly stimulate gastric vagal afferents. To examine this possibility, we recorded the vagal afferent response to close celiac arterial infusion of CCK in the same gastric vagal afferent fibers before and after acute surgical removal of the gastroduodenal junction, including the area containing CCK binding sites. This acute pylorec-

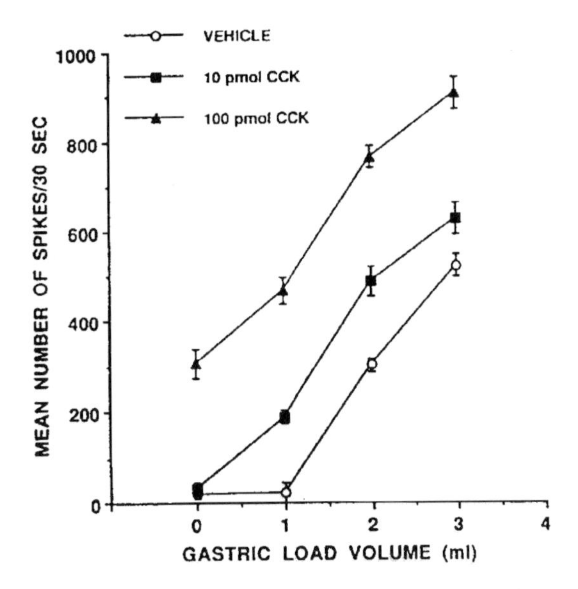

FIGURE 17 Dose-response relationship between gastric saline load volume and neurophysiological spike activity in gastric vagal afferent fibers in the absence (open circles) or in the presence of close celiac arterial infusions of CCK-8 (filled symbols) (Schwartz *et al.*, 1993).

tomy failed to alter the gastric vagal afferent response to CCK infusions, demonstrating that CCK's interactions with this pyloric CCK receptor population were not necessary to mediate the vagal afferent response to CCK (Schwartz *et al.*, 1994). These data do not rule out the possibilities that, during normal digestion, such pyloric activation does lead to gasric vagal afferent excitation or that a vagal afferent response to CCK may be secondary to other local gastric contractile action of the peptide. However, they are consistent with the view that vagal afferent responses to CCK are elicited directly though interactions with vagal membrane bound CCK_A receptors capable of mediating the production of action potentials.

More recent work has demonstrated that CCK also activates vagal afferent fibers with duodenal receptive fields (Schwartz, Tougas, and Moran, 1995). As shown in Figure 19, vagal afferent fibers that respond to duodenal but not to gastric loads also increase their firing to local arterial infusions of CCK. Furthermore, combined application of the duodenal load and close celiac arterial administration of CCK produce a greater degree of activation in these fibers than either stimulus given alone, similar to our findings with gastric mechanoreceptive afferents (Figure 20).

These findings of CCK-induced vagal activation, in combination with the ability of CCK to inhibit gastric emptying, demonstrate multiple potential interactions between CCK and gastric distention. CCK reduces intragastric pressure via a vago-vagal reflex (Raybould, Roberts, and Dockray, 1987), thereby reducing

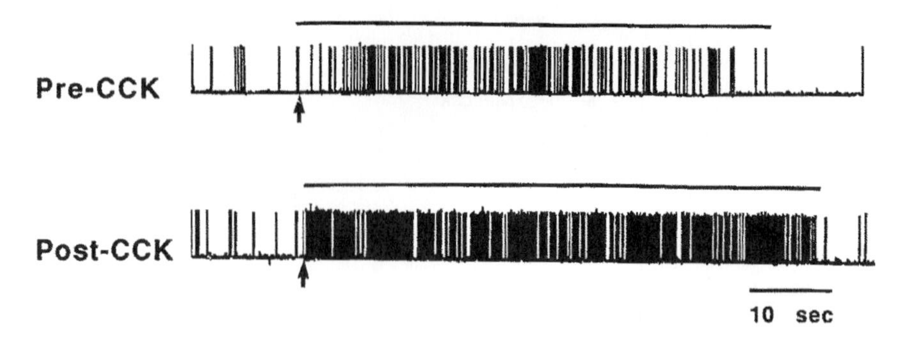

FIGURE 18 Neurophysiological recordings from a representative gastric vagal afferent fiber in response to a 2 ml gastric saline load before (top trace) and after (bottom trace) close celiac arterial infusion of 100 pmol CCK. Arrows indicate onset of stimulus, and solid bars above traces indicate duration of stimulus delivery (Schwartz *et al.*, 1991b).

the pressure gradient between gastric and duodenal compartments critical to slowing the rate of gastric emptying. Thus, CCK itself elicits vagal and, as previously discussed, pyloric events that promote the inhibition of gastric emptying. This multidetermined, CCK-induced inhibition of gastric empyting promotes gastric distention as feeding continues and food accumulates in the stomach. The production of gastric distention, secondary to CCK-induced inhibition of gastric emptying, would in turn activate gastric mechanosensitive vagal afferent fibers. CCK itself mimics the actions of gastric distention in these gastric vagal afferents; the same fibers that respond to gastric distention also respond to CCK. CCK potentiates the response of these vagal afferents to gastric distention; the same degree of distention produces significantly greater vagal activation in the simultaneous or recent presence of CCK. Finally, CCK-induced satiety, CCK-induced inhi-

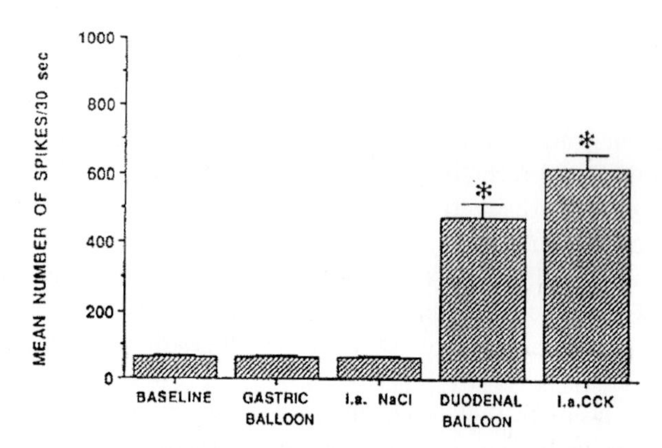

FIGURE 19 Ability of close celiac arterial CCK (100 pmol) but not 2 ml gastric saline loads to stimulate duodenal vagal afferent mechanosensitive fibers (N = 14). Asterisks indicate signifiant difference from baseline condition.

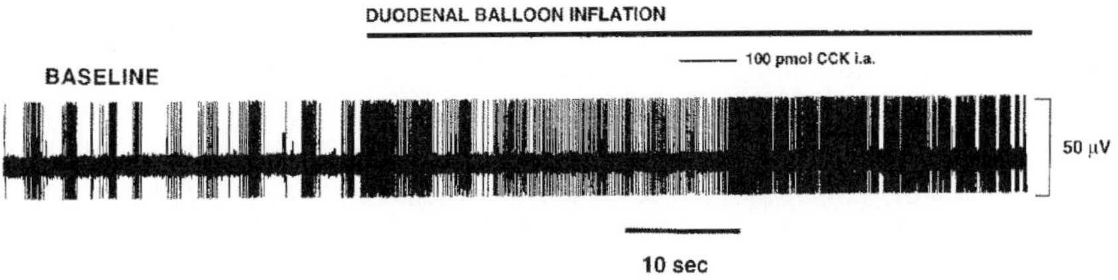

FIGURE 20 Ability of 100 pmol close celiac arterial CCK to enhance the duodenal vagal afferent response to 0.1 ml duodenal balloon inflation. Bar above trace indicates duration of stimulus application (Schwartz *et al.*, 1995).

bition of gastric emptying, and CCK-induced increases in gastric vagal afferent activity all share a common mediation through CCK's actions at the low affinity CCK_A receptor subtype.

Together, results from these neurophysiological studies have revealed that the vagus is not merely a passive conduit, playing a permissive role in the transmission of meal-related signals arising from the gut toward the central nervous system sites mediating the control of food intake. Instead, we have demonstrated that the peripheral afferent vagus nerve supplying gut targets has the capacity to integrate information arising from combinations of meal-related stimuli that are likely to be encountered during real feeding. The degree of integration in these fibers parallels the degree of integration seen between CCK and gastric distention signals in the control of food intake. These findings are paralleled by behavioral and physiological actions of CCK in the inhibition of food intake and gastric emptying, and they are consistent with a role for these integrated vagal signals in providing negative feedback critical to the coordination of gastrointestinal function, the regulation of nutrient delivery to the duodenum, and meal termination.

The pools of endogeous CCK to which these pyloric and vagal elements may be responsive are undetermined. The recent localization of the terminations of gastric and duodenal vagal afferents suggests potential sites of endogenous CCK release that could activate these fibers. Berthoud and Powley (1992) have demonstrated that single vagal afferents innervating the rat stomach terminate in two morphologically distinct types of endings. One type of ending appears as a lattice-like structure parallel with the longitudinal or circular muscle fibers at the site of innervation in the gastric external muscle wall. Terminations such as these would be ideally suited to responding to either local gastric contraction or passive stretch of the local gastric site. These same fibers also have a termination that appears as a calyx-like structure surrounding enteric neurons. These terminals appear to synapse on these enteric neurons. Such a termination would be ideally suited to responding to a local neuronal release of a transmitter substance such as CCK. More recent studies of Berthoud and colleagues (1995) have revealed single vagal afferent fiber endings with terminal arborizations between the crypts and the villous lamina propria of the rat duodenum. These terminations come into close contact with the basal lamina, but they do not actually appear to be in direct contact with the epithelial cells that form the duodenal luminal surface. These duodenal vagal afferent fibers may also come into contact with the myenteric plexus. CCK has been identified in enteric fibers in the duodenum and in the human distal stomach. This may be the pool of CCK to which the gastric and duodenal vagal afferents are responsive.

VII. Perspectives

The potential interactions of signals arising from a variety of sources playing roles in CCK satiety suggest a more general context for the evaluation of signals which contribute to the controls of food intake. Although the majority of early studies of food intake control focused on the contribution of individual neural substrates and/or signals arising from individual gut compartments, more recent work has begun to acknowledge and evaluate the degree to which multiple gastrointestinal and central nervous system events combine and interact to regulate food intake. This approach stems from a physiological perspective that incorporates the distribution and availability of nutrient products of digestion before, during, and at the end of meals.

Feeding terminates while most ingested nutrients are still in the stomach and duodenum, unavailable for metabolic use. This pattern of upper gastrointestinal nutrient distribution relative to the onset of satiety has suggested experiments designed to evaluate where and how the by-products and consequences of ingestion may act *in the periphery* to terminate feeding. Results outlined in this article reveal that the vagus nerve innervating peripheral gut targets is sensitive to meal-related mechanical and peptidergic stimuli, and support the notion that vagal afferent signals play a significant role in the negative feedback control of food intake.

During digestion, nutrients are present at multiple gastrointestinal sites, first sequentially, then simultaneously. This overlapping temporal and spatial pattern of nutrient distribution will stimulate multiple gastric and intestinal sensor elements concurrently as digestion proceeds. Such concurrent stimulation initiates a range of local enteric and long loop (vago-vagal or splanchnic) reflexes that further determine the handling of nutrients, and thus modify the duration of exposure to the mechanical and chemical stimulation provided by the physical and chemical properties of an ingested meal. For example, both luminal exposure to very small volumes of nutrient loads, as well as mechanical distention of the duodenum reflexively slows gastric emptying and promotes gastric fill. Thus, it is likely that during normal ingestion both duodenal and gastric load sensitive vagal afferent would be excited simultaneously as chyme is delivered to the duodenum and as ingesta accumulate in the stomach. In addition, we have demonstrated that the enhanced suppression of food intake produced by combinations of peripheral meal-related gut stimuli is paralleled by the potentiation of gut vagal afferent signals elicited by these combinations. Finally, ingested nutrients stimulate the paracrine and endocrine release of brain–gut peptides. These peptides initiate and modulate

GI signals implicated in the negative feedback control of feeding, in part by coordinating gastrointestinal activity that determines subsequent nutrient delivery.

This summary of recent findings using CCK as a model system for brain–gut communication in the control of food intake reveal potential rules or processes underlying brain–gut communication in general. First, brain–gut peptides may have multiple receptor types, strategically located in peripheral gut sites that determine the handling and/or sensing of nutrients in the gut. Second, brain–gut receptor types may exist in multiple affinity states, and a particular affinity state may mediate a wide range of behavioral, gastrointestinal, and peripheral neural actions of the peptide. Third, brain–gut peptides released by meals may act as stimuli capable of directly generating peripheral neural signals important in terminating food intake. Fourth, brain–gut peptides may modulate the transduction of other meal-elicited negative feedback signals. Finally, brain–gut peptides coordinate gastrointestinal actions that determine nutrient delivery and may themselves produce signals critical to the negative feedback control of food intake.

In conclusion, we view the entire constellation of meal-related stimuli arising from several gastrointestinal compartments as comprising a context representative of the internal milieu during a meal. Negative feedback signals arising from gut contact with meal-related stimuli are but one class of signals to be interpreted in this context. Furthermore, we suggest that the behavioral salience of any individual signal for the termination of food intake is determined by this context. This synthetic perspective demands continued incorporation of a range of behavioral, physiological, neuroanatomical, and pharmacological data from a variety of investigators and laboratories.

Acknowledgments

Supported by DK19302, DK47208, and the Whitehall Foundation.

References

Antin, J., Gibbs, J., Holt, J., Young, R. C., & Smith, G. P. (1975). Cholecystokinin elicits the complete behavior sequence of satiety in rats. *J Comp. Physiol.* **89:** 784–790.

Asin, K. E., Bednarz, L., Nikkel, A. L., Gore, P. A., Jr., Montana, W. E., Cullen, M. J., Shiosaki, K., Craig, R., & Nadzan, A. M. (1992a). Behavioral effects of A-71623, a highly selective CCK-A agonist tetrapeptide. *Am. J. Physiol.* **263:** R125–R135.

Asin, K. E., Gore, P. A., Jr., Bednarz, L., Holladay, M., & Nadzan, A. M. (1992b). Effects of selective CCK receptor agonists on food intake after central or peripheral administration in rats. *Brain Res.* **517:** 169–174.

Bado, A., Durieux, C., Moizo, L., Roques, B. P., & Lewin, M. J. M. (1991). Cholecystokinin-A receptor mediation of food intake in cats. *Amer. J. Physiol.* **260:** R693–R697.

Berthoud, H. R., Kressel, M., Raybould, H. E., & Neuhuber, W. L. (1995). Vagal sensors in the rat duodenal mucosa: Distribution and structure as revealed by in vivo DiI-tracing. *Anat. Embryol.* **191:** 203–212.

Berthoud, H. R., and Powley, T. L. (1992). Vagal afferents innervation of the rat fundic stomach: Morphological characterization of the gastric tension receptor. *J. Comp. Neurol.,* **319:** 261–276.

Brenner, L. A., Yox, D. P., & Ritter, R. C. (1993). Suppression of sham feeding by intestinal nutrients is not correlated with plasma cholecystokinin elevation. *Am. J. Physiol.* **264:** R972–R976.

Buchan, A., Polak, J., Solcia, E., Capella, C., Hudson, D., & Pearse, A. (1978). Electron immunohistochemical evidence for the human intestinal I cell as the source of CCK. *Gut.* **19:** 403–407.

Conover, K. L., Collins, S. M., & Weingarten, H. P. (1988). A comparison of cholecystokinin-induced changes in gastric emptying and feeding in rats. *Amer. J. Physiol.* **255:** R21–R26.

Corp, E. S., McQuade, J., Moran, T. H., & Smith, G. P. (1993). Characterization of type A and type B CCK receptor binding sites in rat vagus nerve. *Brain Res.* **623:** 161–166.

Davison, J. S., & Clarke, G. D. (1988). Mechanical properties and sensitivity to CCK of vagal gastric slowly adapting mechanoreceptors. *Am. J. Physiol.* **255:** G55–G61.

Dourish, C. T., Ruckert, A. C., Tattersall, F. D., & Iversen, S. D. (1989a). Evidence that decreased feeding induced by systemic injection of cholecystokinin is mediated by CCK-A receptors. *Eur. J. Pharmacol.* **173:** 233–234.

Dourish, C. T., Rycroft, W., & Iversen, S. D., Postponement of satiety by blockade of brain cholecystokinin (CCK-B) receptors. *Science,* 1989b; **245:** 1509–1511.

Forster, E. R., Green, T., Elliot, M., Bremner, A., & Dockray, G. J. (1990). Gastric emptying in rats: Role of afferent neurons and cholecystokinin. *Am. J. Physiol.* **258:** G552–G556.

Friedinger, R. M. (1992). Synthesis of non-peptide CCK antagonists. In C. T. Dourish, S. J. Cooper, S. D. Iversen, & L. L. Iversn (Eds.). *Multiple Cholecystokinin Receptors in the CNS* pp. 8–27. Oxford: Oxford University Press.

Fulcrand, P., Rodriguez, M., Galas, M., Lignon, C., Laur, J., Aumelas, A., & Martinez, J. (1988). 2-phenylethylester and 2-pheynlethylamide derivative analogues of the C-terminal hepta- and octapeptide of cholecystokinin. *Int. J. Pept. Protein Res.* **32:** 384–395.

Funakoshi, A., Miyasaka, K., Shinosaki, H., Masuda, M., Kawanami, T., Takata, Y., & Kono, A. (1995). An animal model of congenital defect of gene expression of cholecystokinin (CCK)-A receptor. *Biochem. Biophys. Res. Comm.* **210:** 787–796.

Galas, M.-C., Lignon, M.-F., Rodriguez, M., Mendre, C., Fulcrand, P., Laur, J., & Martinez, J. (1988). Structure-activity relationship studies on cholecystokinin: Analogues with partial agonist activity. *Am. J. Physiol.* **254:** G176–G182.

Garlicki, J., Konturek, P. K., Majka, J., Kwiecien, N., & Konturek, S. J. (1990). Cholecystokinin receptors and vagal nerves in control of food intake in rats. *Am. J. Physiol.* **258:** E40–E45.

Gibbs, J., Young, R. C., & Smith, G. P. (1973a). Cholecystokinin decreases food intake in rats. *J. Comp. Physiol. Psychol.* **84:** 488–495.

Gibbs, J., Young, R. C., & Smith, G. P. (1973b). Cholecystokinin elicits satiety in rats with open gastric fistulas. *Nature.* **245:** 323–325.

Greenberg, D., Torres, N. I., Smith, G. P., & Gibbs, J. (1989). The satiating effect of fats is attenuated by the cholecystokinin antagonist lorglumide. In L. H. Schneider, S. J. Cooper, & K. A. Halmi (Eds.), *The Psychobiology of Human Eating Disorders: Preclinical and Clinical Perspectives* (Vol. 575). (pp. 517–520). New York: NYAS Press.

Grundy, D., Bagaev, V., & Hillsley, K. (1995). Inhibition of gastric mechanoreceptor discharge by cholecystokinin in the rat. *Am. J. Physiol.* **268:** G355–G360.

Hill, R. D., & Woodruff, G. N. (1990). Differentiation of central cholecystokinin receptor binding sites using the non-peptide antagonists MK-329 and L-365,260. *Brain Res.* **526:** 276–283.

Hill, D. R., Singh, L., Boden, P., Pinnocl, G. N., & Hughes, J. (1992). Detection of CCK receptor subtypes in mammalian brain using highly selective non-peptide antagonists. In C. T. Dourish, S. J.

Cooper, S. D. Iversen, & L. L. Iversn (Eds.). *Multiple Cholecystokinin Receptors in the CNS* pp. 57–76. Oxford: Oxford University Press.

Hölzer, H. H., Turkelson, C. M., Solomon, T. E., & Raybould, H. E. (1994). Intestinal lipid inhibits gastric emptying via CCK and a vagal capsaicin-sensitive afferent pathway in rats. *Am. J. Physiol.* **267:** G625–G629.

Honda, T., Wada, E., Battey, J. F., & Wank, S. A. (1993). Differential gene expression of CCK_A and CCK_B receptors in the rat brain. *Mol. Cell. Neurosci.* **4:** 143–154.

Innis, R. B., & Snyder, S. H. (1980). Cholecystokinin receptor binding in brain and pancreas: Regulation of pancreatic binding by cyclic and acyclic guanine nucleotides. *Eur. J. Pharmacol.* **65:** 123.

Kopin, A. S., Lee, Y-M., McBride, E. W., Miller, L. J., Lu, M., Lin, H. Y., Kolakowski, L. F., & Beinborn, M. (1992). Expression cloning and characterization of the canine parietal cell gastrin receptor. *Proc. Natl. Acad. Sci.*, **89:** 3605–3609.

Lee, Y. M., Beinborn, M., McBride, E. W., Lu, M., Kolakowski, L. F., & Kopin, A. S. (1992). The human brain cholecystokinin-B/gastrin receptor. *J. Biol. Chem.* **268:** 8164–8169.

LeSauter, J., Goldberg, J. B., & Geary, N. (1988). CCK inhibits real and sham feeding in gastric vagotomized rats. *Physiol. Behav.* **44:** 527–534.

Liddle, R. A. (1994). Regulation of cholecystokinin synthesis and secretion in rat intestine. *J. Nutrition* **124** (suppl.): 1308S–1314S.

Lin, C. W., and Miller, T. R. (1992). Both CCK-A and CCK-B/gastrin receptors are present on rabbit vagus nerve, *Am. J. Physiol.*, **263:** R591–R595.

Lorenz, D. N., & Goldman, S. A. (1982). Vagal mediation of the cholecystokinin satiety effect in rats. *Physiol. Behav.* **29:** 599–604.

Lorenz, D. N., Kreielsheimer, G., & Smith, G. P. (1979). Effect of cholecystokinin, gastrin, secretin and GIP on sham feeding in the rat. *Physiol. Behav.* **23:** 1065–1072.

Lotti, V. J., & Chang, R. S. L. (1989). A new potent and selective non-peptide gastrin antagonist and brain cholecystokinin receptor (CCK-B) ligand: L-365,260. *Eur. J. Pharmacol.* **162:** 273–280.

Lotti, V. J., Pendelton, R. G., Gould, R. J., Hanson, H. M., Chang, R. S. L., & Clineschmidt, B. V. (1987). In vivo pharmacology of L-364,718, a new potent nonpeptide peripheral cholecsytokinin antagonist. *J. Pharm. Exp. Therap.* **241:** 103–109.

Margolis, R., Moran, T. H., & McHugh, P. R. (1989). *In vitro* response of rat gastrointestinal segments to cholecystokinin and bombesin. *Peptides.* **10:** 157–161.

Melville, L. D., Smith, G. P., & Gibbs, J. (1992). Devazepide antagonizes the inhibitory effect of cholecystokinin on intake in sham-feeding rats. *Pharmacol. Biochem. Behav.* **43:** 975–977.

Mercer, J. G., & Lawrence, C. B. (1992). Selectivity of cholecystokinin receptor antagonists, MK-329 and L-365,260, for axonally transported CCK binding sites in the rat vagus nerve. *Neurosci. Lett.* **137:** 229–231.

Miller, L. J., Holicky, E. L., Ulrich, C. D., & Wieben, E. D. (1995). Abnormal processing of the human cholecystokinin receptor gene in association with gallstones and obesity. *Gastroenterology.* **109:** 1375–1380.

Moran, T. H., Ameglio, P. J., Peyton, H. J., Schwartz, G. J., & McHugh, P. R. (1993a). Blockade of type A, but not type B, CCK receptors postpones satiety in rhesus monkeys. *Am. J. Physiol.* **265:** R620–R624.

Moran, T. H., Ameglio, P. J., Schwartz, G. J., & McHugh, P. R. (1992). Blockade of type A, not type B, CCK receptors attenuated satiety actions of exogenous and endogenous CCK. *Am. J. Physiol.* **262:** R46–R50.

Moran, T. H., Ameglio, P. J., Schwartz, G. J., Peyton, H. J., & McHugh, P. R. (1993b). Endogenous cholecystokinin in the control of gastric emptying of liquid nutrient loads in rhesus monkeys. *Amer. J. Physiol.* **265:** R371–375.

Moran, T. H., Baldessarini, A. R., Salorio, C. F., Lowery, T., & Schwartz, G. J. (1997). Vagal afferent and efferent contributions to the inhibition of food intake by cholecystokinin. *Am. J. Physiol.* **272:** R1245–R1251.

Moran, T. H. Crosby, R. J., & McHugh, P. R. (1991). Effects of pylorectomy on cholecystokinin (CCK) induced inhibition of liquid gastric emptying. *Am. J. Physiol.* **261:** R531–R535.

Moran, T. H., Field, D. G., Carrigan, T. S., and McHugh, P. R. (1995). Role for endogenous cholecystokinin in the gastric emptying of glucose and maltose in rhesus monkeys. *Obesity Res.,* **3:** Suppl. 3: 325s.

Moran, T. H., Kornbluh, R., Moore, K., & Schwartz, G. J. (1994). Cholecystokinin inhibits gastric emptying and contracts the pyloric sphincter in rats by interacting with low affinity CCK receptor sites. *Reg. Peptides.* **52:** 165–172.

Moran, T. H., & McHugh, P. R. (1982). Cholecystokinin suppresses food intake by inhibiting gastric emptying. *Am. J. Physiol.* **242:** R491–R497.

Moran, T. H., & McHugh, P. R. (1988). Gastric and non-gastric mechanisms for satiety action of cholecystokinin. *Amer. J. Physiol.* **254:** R628–R632.

Moran, T. H., Norgren, R., Crosby, R. J., & McHugh, P. R. (1990). Central and peripheral vagal transport of cholecystokinin binding sites occurs in afferent fibers. *Brain Res.* **526:** 95–102.

Moran, T. H., Robinson, P. H., Goldrich, M. S., & McHugh, P. R. (1986). Two brain cholecystokinin receptors: Implications for behavioral actions. *Brain Res.* **362:** 175–179.

Moran, T. H., Shnayder, L., Hostetler, A. M., & McHugh, P. R. (1988). Pylorectomy reduces the satiety action of cholecystokinin. *Am. J. Physiol.* **255:** R1059–R1063.

Moran, T. H., Shnayder, L., Schwartz, G. J., & McHugh, P. R. (1991). Pyloric cholecystokinin receptors. In G. Adler and C. Beglinger (Eds.). *CCK Antagonists in Gastroenterology.* Heidelberg: Springer-Verlag, pp. 159–164.

Moran, T. H., Smith, G. P., Hostetler, A. M., & McHugh, P. R. (1987). Transport of cholecystokinin (CCK) binding sites in subdiaphragmatic vagal branches. *Brain Res.* **415:** 149–152.

Morley, J. E., Levine, A. S., Kneip, J., & Grace, M. (1982). The effect of vagotomy on the satiety effects of neuropeptides and naloxone. *Life Sci.* **30:** 1943–1947.

Murphy, R. B., & Smith, G. P. (1987b). Contractile effects of cholecystokinin-octapeptide (CCK-8) in dispersed rat pyloric sphincter cells. *Appetite, Thirst, and Related Disorders.* 68.

Murphy, R. B., Schneider, L. H., & Smith, G. P. (1988). Peripheral loci for the mediation of cholecystokinin-induced satiety. In R. Wang and R. Schoenfeld (Eds.), *CCK Antagonists* pp. 173–191. New York: Alan Liss.

Murphy, R. B., Smith, G. P., & Gibbs, J. (1987a). Pharmacological examination of cholecystokinin (CCK-8)-induced contractile activity in the rat isolated pylorus. *Peptides* **8:** 127–134.

Muurhainen, N. E., Kissileff, H. A., Lachaussee, J., & Pi-Sunyer, F. X. (1991). Effect of a soup preload on reduction of food intake by cholecystokinin in humans. *Amer. J. Physiol.* **260:** R672–R680.

Passaro, E., Debas, H., Oldendorf, W., & Yamada, T. (1982). Rapid appearance of intraventricularly administered neuropeptides in the peripheral circulation. *Brain Res.* **241:** 338–340.

Raybould, H. R., & Hölzer, H. (1992). Dual capsaicin-sensitive afferent pathways mediate inhibition of gastric emptying in rat induced by intestinal carbohydrate. *Neurosci. Lett.* **141:** 236–238.

Raybould, H. R., Roberts, M. E., & Dockray, G. J. (1987). Reflex decreases in intragastric pressure in response to cholecystokinin in rats. *Am. J. Physiol.* **253:** G165–G170.

Reidelberger, R. D. (1992). Abdominal vagal mediation of the satiety effects of exogenous and endogenous cholecystokinin in rats. *Am. J. Physiol.* **263:** R1354–R1358.

Reidelberger, R. D., & O'Rourke, M. F. (1989). Potent cholecystokinin antagonist L-364,718 stimulates food intake in rats. *Am. J. Physiol.* **257:** R1512–R1518.

Reidelberger, R. D., & Solomon, T. E. (1986). Comparative effects of CCK-8 on feeding, sham feeding, and exocrine pancreatic secretion in rats. *Am. J. Physiol.* **251:** R97–R105.

Reidelberger, R. D., Varga, G., & Solomon, T. E. (1991). Effects of selective cholecystokinin antagonists L 364,718 and L 365,260 on food intake in rats. *Peptides.* **12:** 1215–1221.

Reidelberger, R. D., Varga, G., Liehr, R. M., Castellanos, D. A., Rosenquist, G. L., Wong, H. C., & Walsh, J. H. (1994). Cholecystokinin suppresses food intake by a non-endocrine mechanism in rats. *Amer. J. Physiol.* **267:** R901–R908.

Ritter, R. C., & Ladenheim, E. E. (1985). Capsaicin pretreatment attenuates suppression of food intake by cholecystokinin. *Am. J. Physiol.* **248:** R501–R504.

Sankaran, H., Goldfine, I. D., Bailey, A., Licko, V., & Williams, J. A. (1982). Relationship of cholecystokinin receptor binding to regulation of biological functions in pancreatic acini. *Am. J. Physiol.* **242:** G250–G257.

Sankaran, H., Goldfine, I. D., Deveney, C. W., Wong, K. Y., & Williams, J. A. (1980). Binding of cholecystokinin to high affinity receptors on isolated rat pancreatic acini. *J. Biol. Chem.* **255:** 1849–1853.

Scheurer, U., Varga, L., Drack, E., Burki, H. R., & Halter, F. (1989). Mechanism of action of cholecystokinin octapeptide in rat antrum, pylorus, and duodenum. *Am. J. Physiol.* **257:** R1162–R1168.

Schneider, L. H., Gibbs, J., & Smith, G. P. (1986). Proglumide fails to increase food intake after an ingested preload. *Peptides.* **7:** 135–140.

Schulzenberg, M., Hokfelt, T., Nilsson, G., Terenius, J., Rehfeld, F., Brown, M., Elde, R., Goldstein, M., & Said, S. (1980). Distribution of peptide- and catecholamine-containing neurons in the gastrointestinal tract of rat and guinea pig: Immunohistochemical studies with antisera to substance P, vascoactive inhibitory peptide, enkephalin, somatostatin, gastrin/cholecystokinin, neurotensin, and dopamine ß-hydroxylase. *Neuroscience.* **5:** 689–744.

Schwartz, G. J., & Moran, T. H. (1996). Subdiaphragmatic vagal afferent integration of meal-related gastrointestinal signals. *Neurosci. Biobehav. Rev.* **20:** 47–56.

Schwartz, G. J., McHugh, P. R., & Moran, T. H. (1991b). Integration of vagal afferent responses to gastric loads and cholecystokinin. *Am. J. Physiol.* **261:** R64–R69.

Schwartz, G. J., McHugh, P. R., & Moran, T. H. (1993). Gastric loads and cholecystokinin synergistically stimulate rat gastric vagal afferents. *Am. J. Physiol.* **265:** R872–R876.

Schwartz, G. J., McHugh, P. R., & Moran, T. H. (1994). Pharmacological dissociation of responses to CCK and gastric loads in rat mechanosensitive vagal afferents. *Am. J. Physiol.* **267:** R303–R308.

Schwartz, G. J., Tougas, G., & Moran, T. H. (1995). Integration of vagal afferent response to duodenal loads and exogenous CCK in rats. *Peptides.* **16:** 707–711.

Schwartz, G. J., Moran, T. H., White, W. O., & Ladenheim, E. E. (1997). Relationships between gastric motility and gastric vagal afferent responses to CCK and GRP differ. *Am. J. Physiol.* **272:** R1726–R1733.

Schwartz, G. J., Netterville, L. A., McHugh, P. R., and Moran, T. H. (1991a). Gastric loads potentiate and magnify the inhibition of food intake produced by a cholecystokinin analog. *Am. J. Physiol.* **261:** R1141–R1146.

Shillabeer, G., & Davison, J. S. (1984). The cholecystokinin antagonist, proglumide, increases food intake in the rat. *Regul. Pept.* **48:** 640–641.

Silver, A. J., Flood, J. F., Song, A. M., & Morley, J. E. (1989). Evidence for a physiological role for CCK in the regulation of food intake in mice. *Am. J. Physiol.* **256:** R646–R652.

Smith, G. P., Falasco, J., Moran, T. H., Joyner K. M., and Gibbs, J. (1988). CCK-8 decreases food intake and gastric emptying after pylorectomy and pyloroplasty, *Am. J. Physiol.* **255:** R113–116.

Smith, G. P., Jerome, C., & Norgren, R. (1985). Afferent axons in abdominal vagus mediate satiety effect of cholecystokinin in rats. *Am. J. Physiol.* **249:** R638–R641.

Smith, G. P., Jerome, C., Cushin, B. J., Eterno, R., & Simansky, K. J. (1981). Abdominal vagotomy blocks the satiety effect of cholecystokinin in the rat. *Science.* **213:** 1036–1037.

Smith, G. T., Moran, T. H., Coyle, J. T. Kuhar, M. J., O'Donahue, T. L. & McHugh, P. R. (1984). Anatomic localization of cholecystokinin receptors to the pyloric sphincter. *Am. J. Physiol.* **246:** R127–R130.

South, E. H., & Ritter, R. C. (1988). Capsaicin application to central or peripheral vagal fibers attenuates CCK satiety. *Peptides.* **9:** 601–612.

Wank, S. A., Pisegna, J. R., & de Weerth, A. (1992). Brain and gastrointestinal cholecystokinin receptor family: Structure and functional expression. *Proc. Natl. Acad. Sci.* **89:** 8691–8695.

Wank, S. A., Pisegna, J. R., & de Weerth, A. (1994). Cholecystokinin receptor family. Molecular cloning, structure, and functional expression in rat, guinea pig, and human. In J. R. Reeve, V. Eysselein, T. E. Solomon, & V. L. W. Go (Eds.), *Cholecystokinin.* pp. 49–66. Ann. NY Acad. Sci. (Vol. 713). New York: NYAS Press.

Wank, S. A., Harkins, R., Jensen, R. T., Shapira, H., de Weerth, A., & Slattery,T. (1992). Purification, molecular cloning and functional expression of the cholecystokinin receptor from rat pancreas. *Proc. Natl. Acad. Sci.* **89:** 3125–3129.

Weatherford, S. C., Chiruzzo, F. Y., & Laughton, W. B. (1992). Satiety induced by endogenous and exogenous cholecystokinin is mediated by CCK-A receptors in mice. *Am. J. Physiol.* **262:** R574–R578.

Weatherford, S. C., Laughton, W. B., Salabarria, J., Danho, W., Tilley J. W., Netterville, L. A., Schwartz, G. J., & Moran, T. H. (1993). CCK satiety is differentially mediated by high and low affinity CCK receptors in mice and rats. *Am. J. Physiol.* **264:** R244–R249.

Weller, A., Smith, G. P., & Gibbs, J. (1990). Endogenous cholecystokinin reduces feeding in young rats. *Science.* **247:** 1589–1591.

Yox, D. P., Brenner, L., & Ritter, R. C. (1992). CCK-receptor antagonists attenuate suppression of sham feeding by intestinal nutrients. *Am. J. Physiol.* **262:** R554–R561.

Zarbin, M. A., Innis, R. B., Wamsley, J. K., Snyder, S. H., & Kuhar, M. J. (1983). Autoradiographic localization of cholecystokinin receptors in rodent brain. *J. Neurosci.* **3:** 877–906.

Zarbin, M. A., Wamsley, J. K., Innis, R. B., & Kuhar, M. J. (1981). CCK receptors: Presence and axonal flow in the rat vagus nerve. *Life Sci.* **29:** 697–705.

Zittel, T. T., von Elm, B., Raybould, H. E., Becker, H. D. (1995). Cholecystokinin is partly responsive for food intake and body weight loss after total gastrectomy in rats, *Am. J. Surg.* **169:** 265–270.

PROGRESS IN PSYCHOBIOLOGY AND PHYSIOLOGICAL PSYCHOLOGY, VOL. 17

Fear and Its Neuroendocrine Basis

Jay Schulkin

Department of Physiology and Biophysics
Georgetown University
Behavioral Neuroscience Unit, Clinical Neuroendocrinology Branch
National Institute of Mental Health
Washington, DC 20007

I. Abstract

The premise of this review is that fear is a central motive state that orchestrates both behavioral and autonomic responses to the perception of danger. Like most other central motive states, fear is regulated by neuroendocrine events. One key neuroendocrine mechanism in the adaptation to fear-inducing events is the activation from glucocorticoids of central corticotropin-releasing hormone gene expression in extrahypothalamic sites (central nucleus of the amygdala and the lateral region of the bed nucleus of the stria terminalis) while restraining paraventricular hypothalamic corticotropin-releasing gene expression. This coordinated pattern of gene expression permits the animal to remain vigilant to danger and to curtail the negative impact of sustained hypothalamic-pituitary-adrenal activation.

II. Introduction

Fear is a motivational/emotional state that serves animals in problem solving and in adapting to danger. The emotion of fear is regulated by neuroendocrine events in neural circuits that underlie fear-related behavioral and autonomic responses.

One brain region that is critical in the regulation of fear is the amygdala. I will suggest that one function of glucocorticoid hormones is to facilitate the synthesis of the neuropeptide corticotropin-releasing hormone (CRH) in this nucleus (along with the lateral bed nucleus of the stria terminalis) in maintaining and coping with events that are perceived as frightening. The neuroendocrine hypothesis is that elevated levels of glucocorticoids, secreted by the adrenal gland, act on the amygdala to facilitate CRH gene expression and the central motive state of fear.

I begin with a discussion of the central motive state of fear and its biological basis. I then discuss the neural circuitry that underlies the perception of fearful events. I then move to a description of the neuroendocrine basis of fear and a discussion of glucocorticoids and the CRH in sustaining fear-related behaviors. In each section, I indicate that the same neural and endocrine system underlies pathological states associated with excessive fear. I end with a philosophical discussion of the logical status of the concept of fear in our scientific lexicon.

0363–0951/98 $25.00

III. Central Motive State of Fear

The concept of central motive states historically emerged with those psychologists or biologists investigating the central nervous system and the organization of behavior. The term *central motive state* was coined by Lashley (1938) and was quickly embraced by Beach (1942) or what Beach called "central excitatory states." It was taken up by a number of ethologists (e.g., Tinbergen, 1951), psychobiologists (e.g., Hebb, 1949), and physiological psychologists (e.g., Morgan and Stellar, 1950).

The concept of central motive states anchored to basic biological systems found a clear use in neuroscience and in understanding behavior, measuring, for example, the performance of animals when they are hungry, thirsty, sexual, or fearful (see Gallistel, 1980; Pfaff, 1980). The classic paper by Eliot Stellar (1954) titled "The Physiology of Motivation" along with the elegant experimental designs of Neil Miller (1957) legitimated the use of the term in modern scientific vernacular. Central motive states, such as fear, were placed in functional biological contexts. It was therefore understood that the central nervous system generates behavioral responses. The central nervous system embodies, for example, the central state of fear and its adaptive role in motivating behavioral responses.

Fear is a prototypical exemplar of a central state—a state of the brain. Although systemic physiological changes influence the state of fear (James, 1884; 1890), peripheral changes are not sufficient for the emotional expression of motivated behaviors such as fear (Bard, 1939; Cannon, 1915). It is the change in the brain that is linked to the state of fear. We are afraid because we perceive danger (see e.g., Aristotle). But bodily events influence and reinforce the state (James, 1890; Damasio, 1996). For example, changes in heart rate, blood pressure, respiration, facial muscles, and catecholamines—both peripheral and central—(e.g., Yang, Gorman, and Dunn, 1990) influence the state of fear (see LeDoux, 1996).

The central state of fear is knotted to attention and learning (Mackintosh, 1975) and the assessment of relevant information (Dickinson, 1980), which are important in predicting future outcomes (Miller, 1959; Rescorla and Wagner, 1972). The central state of fear is tied to action tendencies (Frijda, 1986), attention (Lang, 1995), and appraisals more generally of environmental stimuli (LeDoux, 1996).

Fear is linked to an appetitive system (which includes consummatory behaviors) and an aversive system (e.g., withdrawal, protectiveness, and so on). The former in Konorski's (1967) terms is preservative and the later is protective. That is, fear maps on to approach/appetitive and avoidance/withdrawal mechanisms influenced by sensory stimulation (Schnierla, 1959), which Konorski (1967) characterized. Both approach and avoidance behavioral responses are represented in the brain (e.g., Konorski, 1967; Gray, 1971, 1982). For example, both laboratory and clinical observations have suggested the hypothesis that the left prefrontal cortex is linked to approach responses and the right prefrontal cortex is linked to avoidance or withdrawal responses (Davidson, 1994).

IV. Psycholobiological Basis of Fear

Fear is an adaptive response in the perception of danger, and it is fundamental in problem solving and longevity. In fact, fear as an emotion evolved as part of problem solving (Darwin, 1872). Emotions like fear prepare the animal to respond to danger by heightening vigilant attention (Gallagher and Holland, 1994; Lang, 1995) and motivating behavior (Mowrer, 1947; Bindra, 1978) that includes defensive behaviors (Booles and Fanselow, 1980). The state of fear is one in which there is a readiness to perceive events as dangerous or alarming (Rosen *et al.*, 1996; Le Doux, 1987). The state is knotted to learning about what is safe and what is not (Miller, 1959) and the informational value of stimuli that has predictive value to the animal (Rescorla and Wagner, 1972; Dickinson, 1980).

The physiological and behavioral responses aroused by fear are rooted in our evolutionary past. They are adaptive (Lazarus, 1991); but fear is certainly not always in our everyday life. The traditional view of emotions, such as fear, characterized them as linked to dysfunction (e.g., Spinoza, 1677; Freud, 1926; Goldstein, 1939; Sabini and Silver, 1996), disorganization (Hebb, 1949), or magical thinking (Sartre, 1948). The emotions or passions, as they were historically construed rendered one passive. The root of passion is to be passive, reason by contrast is active (e.g., Spinoza, 1677). The pathological expression of normal fear leads to terror and perhaps the sense of immobilization that many theorists have suggested (Darwin, 1872). But this strikes me as the extreme; terror after all is an extreme state. And under these conditions the adaptive part of fear has reached its limit; pathology now sets in (Rosen *et al.*, 1996).

Once one roots emotions, such as fear, as part of the active and fast response to danger, they are quite rational and organize behavior to avoid harm. Moreover, the emotions are cognitive; appraisals of danger are cognitive. An assessment can be fast and still cognitive (Chomsky, 1972); after all language is fast and is paradigmatic of what we mean by cognitive (Fodor, 1981; 1983). Moreover, the preparedness to associate events as dangerous (snakes, heights, and so forth) may be biologically prewired (Mineka, 1986; Ohman, 1993) and still cognitive. Emotions are cognitive because appraisals are involved (e.g., Arnold, 1960). Something need not be rational to be cognitive; certainly fear is not always rational. However, evolution selected an "affective computational" system that underlies states such as fear (LeDoux, 1993; Parrott and Schulkin, 1993).

Thus emotions like fear are linked to action tendencies (Frijda, 1986) and motivate behavior in response to danger (Ohman, 1993). Fear prepares the animal to freeze or flee (Blanchard and Blanchard, 1972; Booles and Fanselow, 1980). Behaviors such as startle and freezing are expressions of fear across many species.

FIGURE 1 Depiction of a man in terror (Darwin, 1872).

Fear is linked to defensive behavior; however, they are not the same. Fear functions to alert the animal to danger. The motivated animal seeks relief from this state, and with the elimination of fear, there is the sense of relief (Miller, 1959; Mowrer, 1947). In other words, fear functions as a central motive state in threatening contexts resulting in defensive behaviors that are adaptive in reducing or warding off harm. But the perception of fearful events may be constrained by neuronal processing of information. The vigilance that is required during fear limits the attentional mechanisms that might normally be used elsewhere (Davis *et al.,* 1993).

Fear is also a communicative device to others in an elaborate social orchestration (Smith, 1977; Marler and Hamilton, 1966; Ekman and Davidson, 1994). Submissive behaviors, facial displays, and acoustic signals are part of the central state of fear (Hauser, 1996). The central state of fear embodies an elaborate and complex organization of behavior that includes social signals that reduce fighting and maintain alliances (Marler and Hamilton, 1966; Hauser, 1996).

Finally, there is more than one kind of fear (Hebb, 1946a; b; Kagan and Schulkin, 1995); fear of unfamiliar events is not the same as conditioned fear. For example, Hebb in his early studies emphasized fear of unfamiliar objects or of familiar objects that are perceptually deranged (Hebb, 1946a; b; 1949). Discrepant events not only elicit fear-related learning, but other forms of learning (Dewey, 1894, 1895; Rescorla and Wagner, 1972; Dickinson, 1980).

V. Neural Circuits Mediating Fear: Importance of the Amygdala

The amygdala is centered in the temporal region of mammals (Herrick, 1905). It is almond shaped and was originally called the “smell brain.” It has long been considered part of the limbic system in the organization of emotional responses (e.g., Bard, 1939; Papez, 1937).

Regions of the amygdala have been characterized as a "sensory gateway" (Aggleton and Mishkin, 1986; LeDoux *et al.,* 1990; Turner and Herkenham, 1992) because its receives information from both cortical and subcortical regions (Krettek and Price, 1978; Amaral *et al.,* 1992). Specifically, the lateral and basal lateral regions are richly innervated from neocortical and subcortical sites, which relay this information to the central nucleus (e.g., Krettek and Price, 1978; Turner and Herkenham, 1992). The central nucleus also receives visceral information from brainstem sites that include the solitary and parabrachial nuclei (Ricardo and Koh, 1978; Norgren, 1976) and reciprocally project to these brainstem regions (e.g., Schwaber *et al.,* 1982). The amygdala's direct link to the nucleus acumbens led Nauta (1961; Nauta and Domesick, 1982) to suggest an anatomical route by which motivation and motor control action are linked in the organization of behavior (see also Mogenson, 1987).

Damage to the amygdala interferes with fear-related behavioral responses (e.g., Fonberg, 1972; Kling, 1981). In the past 10 years, evidence has converged to show particularly that the central nucleus within the amygdala orchestrates the behavioral responses (LeDoux, 1995; 1996). Lesions, or stimulation, of the central nucleus are known to influence behaviors associated with fear (Kapp *et al.,* 1979; LeDoux *et al.,* 1988). Stimulation of the central nucleus of the amygdala, for example, activates the neural circuitry underlying startle responses and amplifies the startle response (Rosen and Davis, 1988). Stimulation of the amygdala heightens attention toward events that are perceived as fearful (Gallagher and Holland, 1994; Rosen *et al.,* 1996). Or put differently, amygdala activation increases the likelihood that an event will be perceived as fearful. Amygdala activation has been linked to attention to uncertain and aversive events (Gallagher and Holland, 1994) and to anticipatory angst (Schulkin, McEwen, and Gold, 1994). Infusions into the central, or lateral, nuclei of N-methyl-D-aspartase (NMDA) antagonists interferes with fear-related conditioning (Fanselow, 1994; Davis *et al.,* 1993). Neurons within the amygdala are reactive to fearful signals (Armony *et al.,* 1995). The amygdala and its associated neural circuitry appraise fearful signals and orchestrate behavioral and autonomic responses to these events.

In elegant detail LeDoux and his colleagues (e.g., LeDoux *et al.,* 1988; LeDoux, 1995; 1996) have outlined an anatomical circuit underlying conditioned freezing in rats to an auditory cue. In part, it consists of pathways from the medial geniculate nucleus en route to the lateral and central nuclei of the amygdala. In addition, projections from the auditory and perirhinal regions of the neocortex through the lateral nucleus en route to the central nucleus of the amygdala convey information about the acoustic conditioning. Interruption of this input to the lateral and central nucleus of the amygdala impairs the fear conditioning. The central nucleus through its projections to the central gray regulates freezing and escape behaviors (LeDoux, 1987; 1996).

The neural circuitry for both conditioned freezing, outlined previously, and for conditioned startle (Rosen *et al.,* 1991) or for unconditioned fear (Fanselow, 1994), as I indicated, requires the lateral region of the amygdala to receive information and the central region to orchestrate the behavioral and autonomic responses. Other regions in the forebrain organizing fear include the prefrontal cortex (Morgan and LeDoux, 1995), the perirhinal cortex (Rosen *et al.,* 1992), the hippocampus (Kim *et al.,* 1993; Phillips and LeDoux, 1992), and the bed nucleus of the stria terminalis, which as I will describe may be linked to the neuroendocrine regulation of fear (see below).

In an experiment with Jeff Rosen and colleagues (1996), we looked at the effects of amygdala kindling on fear conditioning. Kindling is a way in which to excite the brain through electrodes targeted to an anatomical site and electrical current delivered to the brain region (see e.g., Adamec, 1990). In other words, the result of this experimental manipulation is a putative hyperexcitable site in the brain. The idea was that partial amygdala kindling would potentiate fear-related behavioral responses but that dorsal hippocampal kindling would not. Rats were conditioned to be fearful to light paired with foot shock. During the next several days they received partial kindling of the amygdala or the dorsal hippocampus. Rats were then presented with an auditory startle stimulus with or without light. Fear-potentiated startle was tested one week later (Rosen *et al.,* 1996). The group that underwent amygdala kindling displayed elevated startle amplitude; this did not occur in the group that received hippocampal kindling. See Figure 2.

There is also a good deal of evidence in humans that the amygdala is linked to fear (see LeDoux, 1996). For example, recently, it has been observed that lesions of the amygdala impair fear-related behavior and autonomic responses to conditioned stimuli (e.g., LaBar *et al.,* 1995; Angrilli *et al.,* 1996; Bechara *et al.,* 1993). Several studies have found that lesions of the amygdala interfere with the recognition of fearful facial expression (Adolphs *et al.,* 1995; Allman *et al.,* 1994). Moreover, positron emission tomography (PET) imaging studies have shown greater activation of the amygdala during fear and anxiety provoking stimuli (Ketter *et al.,* 1996). Such PET studies have revealed that the amygdala is activated when presented with fearful as opposed to happy faces (Morris *et al.,* 1996). With the use of functional magnetic resonance imaging (MRI), it has further been shown that the amygdala is activated and then habituates when shown fearful in contrast to neutral or happy faces (Breiter *et al.,* 1996). See Figure 3.

Clinically, some forms of depression (melancholic) are associated with fear (Goodwin and Jamison, 1990; Gold, Goodwin, and Chrousos, 1988). Elevated blood flow to the amygdala has been observed using PET in patients who are fearful and depressed (Drevets *et al.,* 1992). The metabolic rate of the amygdala, in humans, has also been used to both predict depression and negative affect (Abercrombie *et al.,* in press; Ketter *et al.,* 1996).

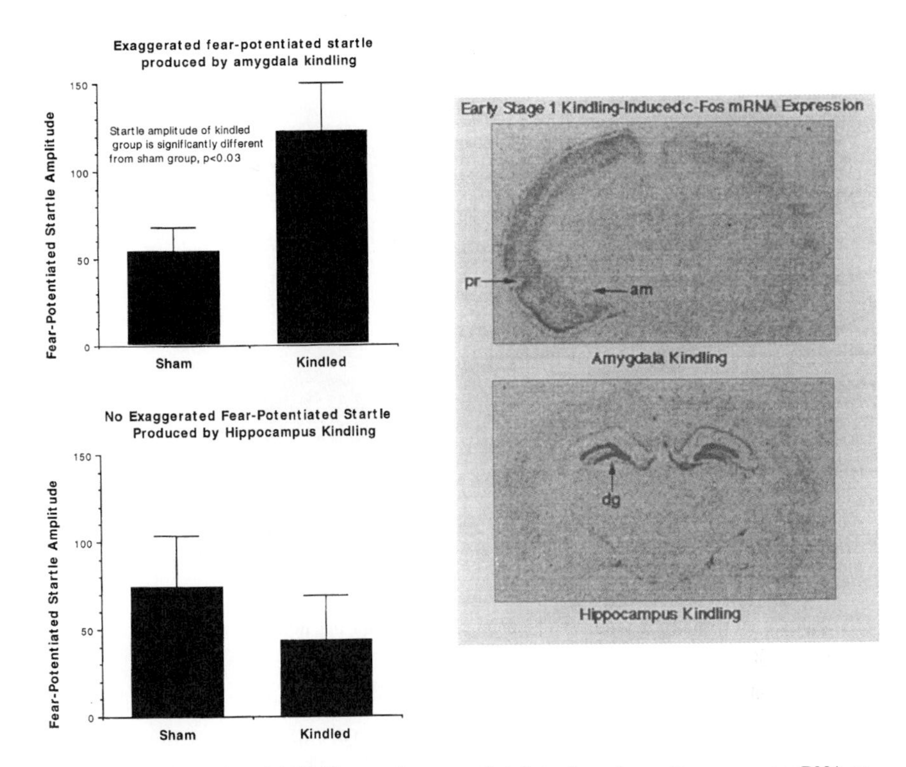

FIGURE 2 Effects of partial kindling on fear-potentiated startle and on c-Fos messenger RNA expression at the level of the amygdala (am) and hippocampus. Pr: perirhinal cortex; dg: dentate gyrus (Rosen *et al.*, 1996).

VI. Glucocorticoids and Corticotropin-Releasing Hormone in the Regulation of Fear

Fear is sustained by neuroendocrine events. Under duress, the hypothalamic-pituitary-adrenal (HPA) axis is activated (e.g., Cannon, 1915; Selye, 1956), as are sites in the brain that participate in the regulation of fear. Consider the glucocorticoids first. The secretion of glucocorticoids helps sustain a number of behavioral responses including fear-related behaviors (Richter, 1949). Without glucocorticoids, as Richter (1949) noted, animals die under duress (see also Selye, 1956). Adrenalectomized animals are unable to tolerate fear, duress, or chronic stress and suffer fatality. Glucocorticoids prepare the animal to cope with emergency and taxing environmental contexts (Cannon, 1915; Richter, 1949).

Glucocorticoids are also essential in the development of fear (Takahashi, 1994; 1995). Removal of corticosterone in rats before, but not after, 14 days of age impairs fear of unfamiliar objects (Takahashi and Rubin, 1993). In other words, there is a critical period in neonatal development in which glucocorticoids facilitate the normal expression of fear of unfamiliar objects (Takahashi, 1994).

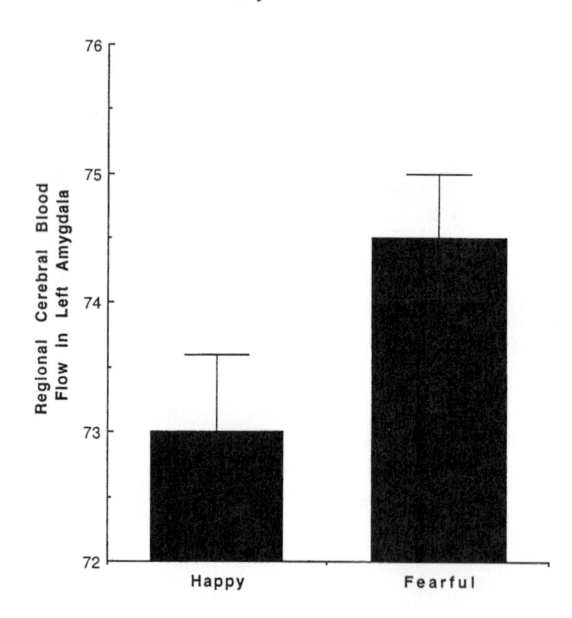

FIGURE 3 Differences in regional blood flow (ml/deciliter per min) in the human left amygdala while viewing happy or fearful faces (Morris *et al.*, 1996).

Glucocorticoids are secreted under a number of experimental conditions in which fear, anxiety, novelty, and uncertainty are experimental manipulations (Mason, Brady, and Sidman, 1957; Mason, 1975; Brier *et al.*, 1987). In contexts in which there is loss of control, or the perception of it, and worry is associated with the loss of control, glucocorticoids are secreted. This holds across a number of species, including humans (e.g., Brier *et al.*, 1987). (See Figure 4.) Perceived control reduces the levels of glucocorticoids that circulate. In rats, for example, predicting the onset of an aversive signal reduces the level of circulating glucocorticoids (Kant *et al.*, 1992). And on the clinical side, one of the most consistent findings in fearful, depressed patients is elevated levels of cortisol and an enlarged adrenal cortex (Sachar *et al.*, 1970; Carroll *et al.*, 1976; Nemeroff *et al.*, 1992). These findings are congruent with those of Richter (1949) who observed an enlarged adrenal gland in stressed, fearful wild rats when compared to unstressed, laboratory analogs.

From a biological view the chronic activation of glucocorticoid hormones is costly. The subordinate male macaque, for example, has elevated cortisol levels but lower levels of testosterone than the dominant one (Sapolsky, 1992). The lower level of testosterone decreases its reproductive fitness. The cost of chronic subordination is perhaps more fearfulness and uncertainty of attack as well as a decrease in the likelihood of successful reproduction. This phenomenon of high corticosterone and low testosterone has been demonstrated in a number of species (see e.g., Lance and Elsey, 1986).

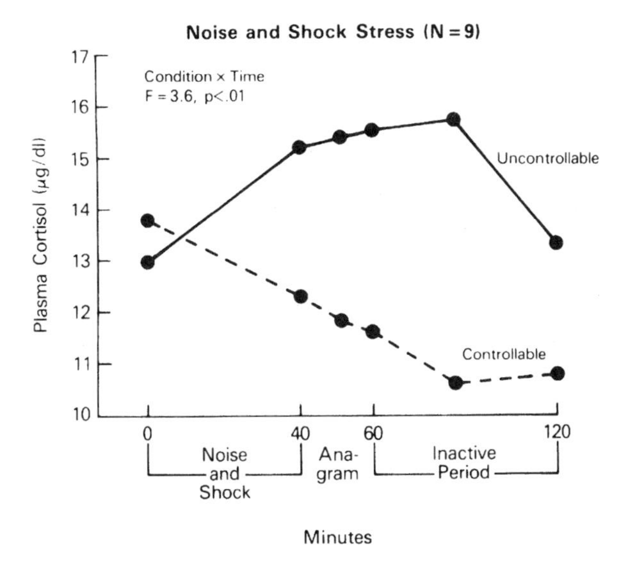

FIGURE 4 Plasma cortisol levels in two uncontrollable–controllable stress paradigms in humans (Brier, 1989).

Fear is also a metabolically costly event. Although glucocorticoids are essential in the development of neuronal tissue and in adapting to duress (Gould, 1994), to sustain this response over time, tissue will begin to deteriorate (Sapolsky, 1992; McEwen *et al.,* 1993). Chronic glucocorticoid activation, for example, increases the likelihood of neurotoxicity and neural endangerment through the loss of glucocorticoid receptors.

Perhaps, to avoid this deterioration, negative regulation of the HPA axis evolved to restrain the stress response. In other words, glucocorticoids restrain the output of the paraventricular nucleus of the hypothalamus (PVN) and pituitary gland, decreasing CRH and adrenocorticotropic hormone (ACTH), and thereby limiting their own production (Munck, Guyre, and Holbrook, 1984; Dallman *et al.,* 1993; Sawchenko, 1987). This is classical negative feedback. Negative feedback is one mechanism to restrain the activation of the HPA (Munck, Guyre, and Holbrook, 1984; see e.g., Figure 9). The restraint of CRH at the level of the PVN is profound and sustained over time. The restraint of HPA function appears to be regulated in part through glucocorticoid activation of the hippocampus and bed nucleus of the stria terminalis (Sapolsky, Zola-Morgan, and Squire, 1991; Cullinan, Herman, and Watson, 1993; Beaulieu *et al.,* 1987). This occurs in part through efferent control of the PVN by gamma-aminobutyric acid (GABA)-mediated inhibitory neurons (Herman and Cullinan, 1997).

Consider now CRH. Corticotropin-releasing hormone is a peptide hormone initially characterized in the PVN. It is a 41-amino acid peptide hormone that facilitates the secretion of ACTH (Vale *et al.,* 1981). But CRH is also synthesized in a number of sites outside the PVN. They include the central nucleus of the amygdala and the lateral bed nucleus of the stria terminalis (Swanson *et al.,* 1983; Gray, 1990; Ju, Simerly, and Swanson, 1989). Corticotropin-releasing hormone is also synthesized in the prefrontal cortex and in cells around the locus ceruleus in Barrington's nucleus, parabrachial region, and solitary nucleus. Moreover, fiber pathways connect many of these regions with one another (Swanson *et al.,* 1983; Gray, 1990). Retrograde labeling and lesion studies, for example, have demonstrated that CRH neurons in the central nucleus and bed nucleus of the stria terminalis are reciprocally connected with one another, along with projections to the central gray, locus coeruleus, and parabrachial nucleus (Gray, 1990; Sawchenko, 1993). Moreover, many of these CRH-producing cells are colocalized with glucocorticoid receptors (e.g., Honkaniemi *et al.,* 1992; Kainu *et al.,* 1993). See Figure 5.

A recently cloned CRH receptor contains a 451-amino acid protein and is linked to a G protein and adenylyl cyclase to increased intracellular cyclic adenosine monophosphate (cAMP) and calcium levels (Owens and Nemeroff, 1991). More recent studies have revealed that there are several different CRH-receptor subtypes, one of which is dominant in limbic regions, and one of which is more widespread (Lovenberg *et al.,* 1996).

It is well known that central infusion of CRH, and not peripheral infusion, potentiates fear of unfamiliar events, startle, or freezing behaviors (e.g., Britton *et al.,* 1986; Koob *et al.,* 1993; Swerdlow, Britton, and Koob, 1989). This has been demonstrated in a number of species and across a number of fear-related behaviors (see Kalin, 1985; Koob and Bloom, 1985). One clear result from CRH infu-

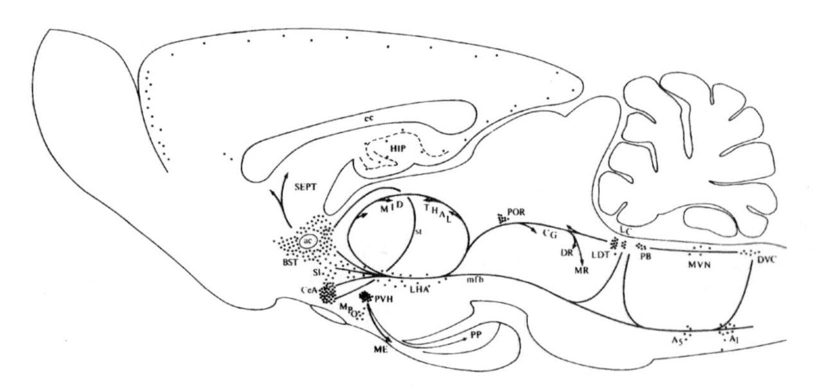

FIGURE 5 Major corticotropin-releasing factor-stained cell groups (dots) and fiber systems in the rat brain. cc, corpus callosum; HIP, hippocampus; SEPT, septal region; AC, anterior commissure; BST, bed nucleus of the stria terminalis; ST, substantia innominata; CcA, central nucleus of the amygdala; MPO, medial preoptic area; PVH, paraventricular nucleus of hypothalamus; ME, median eminence; PP, posterior pituitary; LHA, lateral hypothalamic area; mfb, medial forebrain bundle; MID THAL, midline thalamic nuclei; ST, stria terminalis; POR, perioculomotor nucleus; CG, central gray; DR, dorsal raphe; MR, medial raphe; LDT, laterodorsal tegmental nucleus; LC, locus ceruleus; PB, parabrachial nucleus; MVN, medial vestibular nucleus; DVC, dorsal vagal complex; A^1A^5, noradrenergic cell group (Swanson *et al.,* 1983).

sions is a set of coordinated behavioral and physiological responses that includes enhanced arousal and fear, as well as inhibition of vegetative functions and sexual behavior (e.g., Gold, Goodwin, and Chrousos, 1988; Koob and Bloom, 1985). Corticotropin-releasing hormone injections directly into the central nucleus of the amygdala increase fear related to unfamiliar environments and increase freezing in anticipation of aversive events (Wiersma *et al.,* 1993). Importantly, lesions of the central nucleus of the amygdala, and not the PVN, impair CRH-induced effects on conditioned fear-related startle responses (Liang *et al.,* 1992). Interference of CRH expression by pharmacological antagonists within the central nucleus of the amygdala disrupts fear-related behavioral responses (Koob *et al.,* 1993).

Pathophysiologically, one of the first outstanding observations about CRH was that it was elevated in the cerebrospinal fluid of melancholic, depressed patients (Nemeroff *et al.,* 1984; Gold *et al.,* 1984; Holsboer *et al.,* 1984). See Figure 6. These patients are chronically replete with anticipatory angst. Their appetite for food and sex is reduced, and their sleep and immune responses can be compromised (Gold, Goodwin, and Chrousos, 1988).

Clinicians interpreted this situation of elevated systemic levels of cortisol and high central CRH as an aberration and a reflection of the pathology because high cortisol should be associated with low central CRH. However, an alternative interpretation was that the high cortisol was activating CRH gene expression within specific regions of the brain, resulting in greater fear (Schulkin, 1994; see the following discussion).

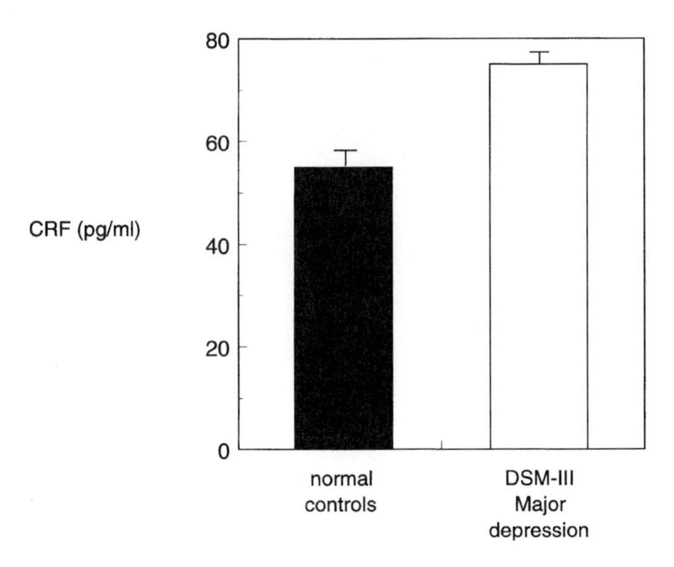

FIGURE 6 Concentration of corticotropin-releasing factor (CRF) in the cerebrospinal fluid of depressed and fearful patients and in normal controls (Nemeroff *et al.,* 1984).

VII. Corticotropin-Releasing Hormone Gene Expression and Function in Extrahypothalamic Sites

We thus began a line of inquiry with the hypothesis that the regulation of hypothalamic CRH gene expression and its restraint by glucocorticoids was one system linked to systemic regulation, and a second CRH system in extrahypothalamic sites, such as the central nucleus of the amygdala and lateral bed nucleus, that we thought might be regulated quite differently and linked to the central state of fear.

We treated adrenally intact rats with systemic high levels of corticosterone either over 4 days, or over a 1-week or a 2-week period (Makino, Gold, and Schulkin, 1994a). The results from one study are shown in Figures 7, 8, and 9. Our results demonstrated that high levels of corticosterone can result in elevated levels of CRH messenger (mRNA) in the central nucleus of the amygdala, particularly in the centromedial region of the central nucleus. This is very much in contrast to the reduction of CRH mRNA in the PVN in the same animals.

As we were completing the work, we discovered that Swanson and Simmons (1989), in several adrenalectomized rats, and Watts and Sanchez-Watts (1995) had found something similar in differentiating the effects of corticosterone on CRH gene

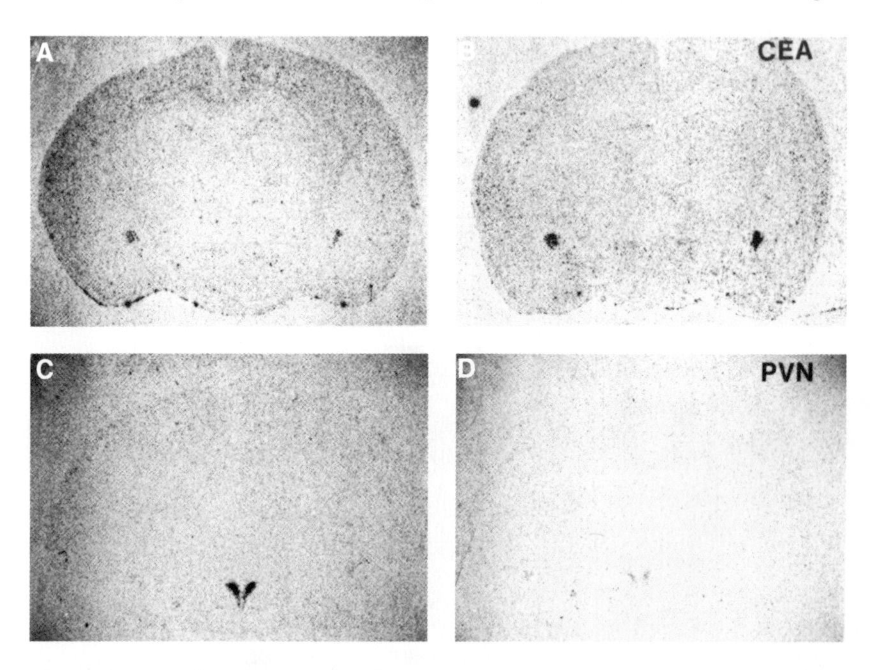

FIGURE 7 Autoradiographs depict the effects of corticosterone (CORT) injection on CRH mRNA levels of the central nucleus of the amygdala (CEA; A is the control and B is the CORT treated) and the paraventricular nucleus of the hypothalamus (PVN; C is the control and D is the CORT treated). Black silver grains mark the location of hybridized probe, and accumulation of grains are seen in the CEA and PVN (Makino, Gold, and Schulkin, 1994a).

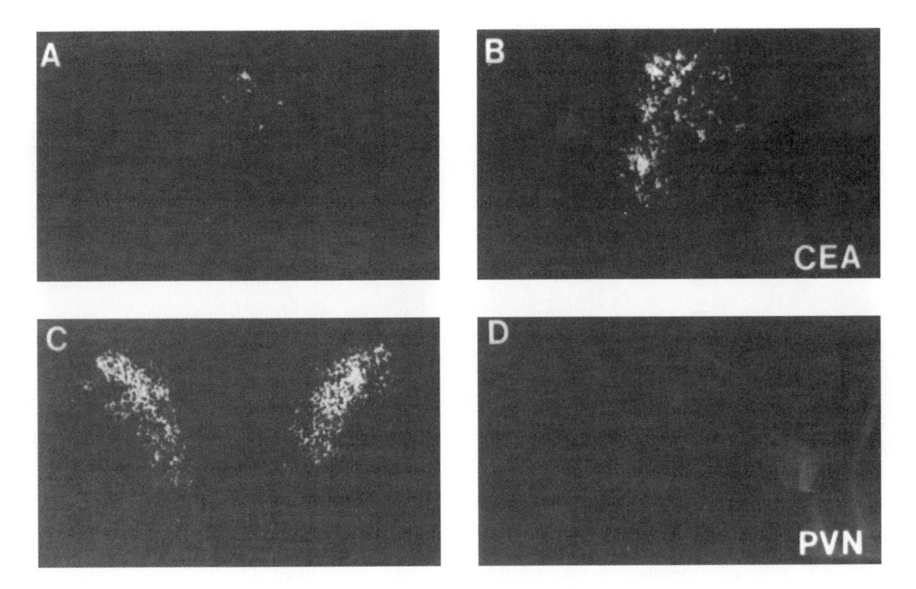

FIGURE 8 Effects of corticosterone (CORT) pellet implanatation on CRH mRNA levels in the central nucleus of the amygdala (CEA) and the paraventricular nucleus (PVN). Dark-field photomicrographs show the autoradiographic distribution of CRH mRNA in the CEA (A and B) and in the PVN (C and D). Controls are depicted in A and C, and CORT treated are depicted in B and D. Black silver grains mark the location of hybridized probe, and accumulation of grains are seen in the CEA and the PVN (Makino, Gold, and Schulkin, 1994a).

expression in the central nucleus of the amygdala from that of the PVN. This effect is clearly shown in Figure 10 in which CRH mRNA is decreased in the PVN while it is increased simultaneously in the central nucleus of the amygdala (Watts and Sanchez-Watts, 1995; Watts, 1996). Thus there is differential regulation of CRH gene expression at the level of the PVN and the central nucleus of the amygdala.

In addition, other evidence has demonstrated that CRH is elevated in the central nucleus of the amygdala during restraint stress—a condition of adversity and presumably fear. Under these conditions corticosterone is also elevated. One study using microdialysis to measure CRH levels found that CRH was elevated during restraint stress in the central nucleus of the amygdala (Pich *et al.,* 1993; 1995). And in another study, under similar experimental conditions, the experimenters found increased CRH mRNA levels measured by *in situ* histochemistry (Kalin, Takahashi, and Chen, 1994).

We then extended our studies to include the bed nucleus of the stria terminalis, which has been considered "extended amygdala" (Johnston, 1923; for a description of the extended amygdala, see Alheid and Heimer, 1988; and Alheid, deOlmos, and Beltramino, 1996; but also Canteras, Simerly, and Swanson, 1995; Petrovich and Swanson, 1997). Both the bed nucleus and the central nucleus contain a number of neuropeptide-producing cells, including CRH (Gray, 1990;

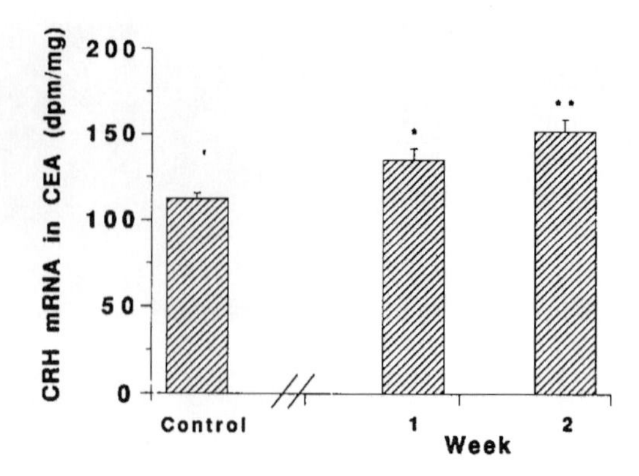

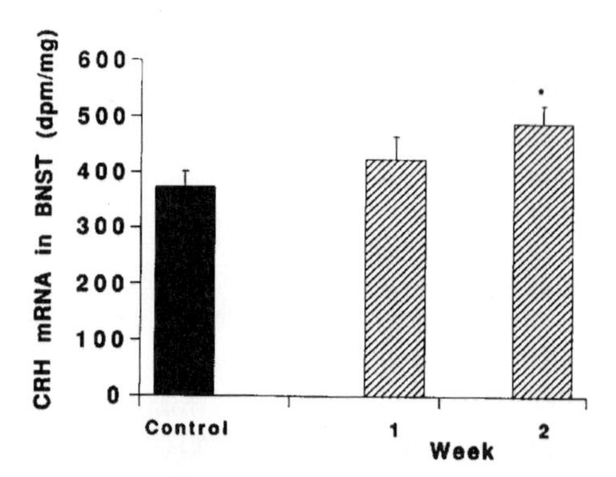

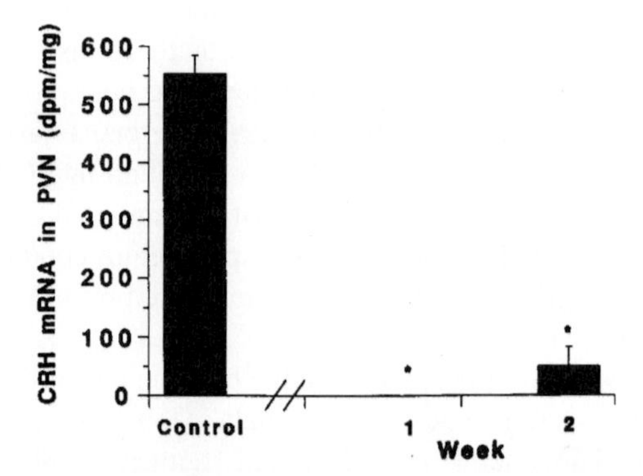

FIGURE 9 CRH mRNA hybridization levels in the central nucleus of the amygdala (CEA), bed nucleus of the stria terminalis (BNST), and paraventricular nucleus of the hypothalamus (PVN) following corticosterone treatment(* = statistical significance) (Makino, Gold, and Schulkin, 1994a and b).

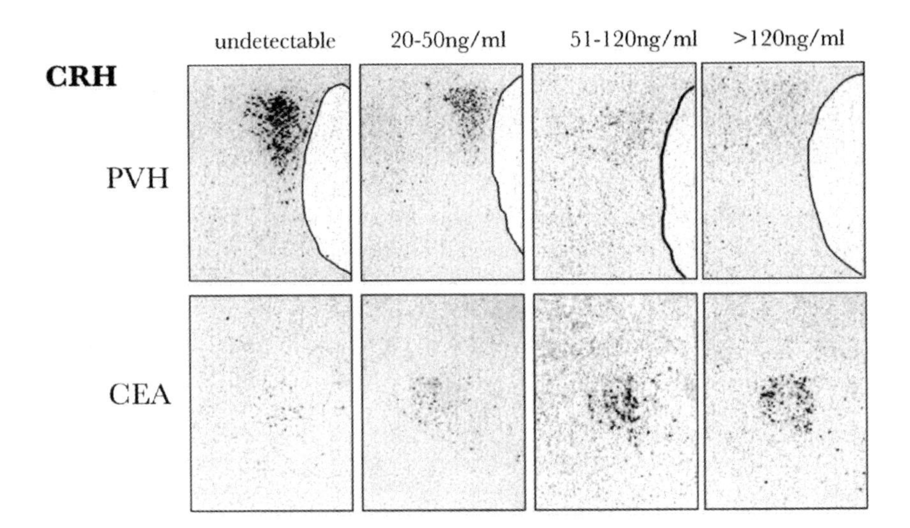

FIGURE 10 Dark-field photomicrographs of CRH mRNA hybridization in the paraventricular nucleus (PVH) and central nucleus (CEA) after increasing plasma corticosterone concentration for 5 days (data from Watts and Sanchez-Watts, 1995).

Mogga *et al.,* 1992). The lateral region of the bed nucleus corresponds with that of the central nucleus, and it is this region that is rich in CRH-producing cells, particularly the dorsal lateral region (Makino, Gold, and Schulkin, 1994b; Watts and Sanchez-Watts, 1995; Gray, 1990; Ju, Simerly, and Swanson, 1989). We again found that CRH mRNA was increased following corticosterone pretreatment in this region (Makino, Gold, and Schulkin, 1994b). Similar effects were also observed in adrenalectomized rats treated with corticosterone on CRH mRNA in the dorsal lateral bed nucleus (Watts and Sanchez-Watts, 1995).[1] See Figures 11 and 12.

In addition, there is some evidence to suggest differential regulation of CRH mRNA from that of the PVN at the level of Barrington's nucleus (Pavcovich and Valentino, 1997), which lies ventromedial to the locus coeruleus (Valentino *et al.,* 1992). Infusions of CRH or antagonists within this region influence fear-related behavioral responses (Swiergiel *et al.,* 1992). Recall that this region is innervated

[1]The receptors for CRH are located primarily outside the central nucleus of the amygdala or the lateral bed nucleus of the stria terminalis. They are present in the lateral nucleus and the medial nucleus of the amygdala (e.g., Makino *et al.,* 1995). It is also now known that there are several CRH receptor sites that may be functionally differentiated (e.g., Lovenberg *et al.,* 1996). The regulation of CRH receptors in these extrahypothalamic sites is unclear (DeSouza *et al.,* 1987; Perrin *et al.,* 1993). However, we did find a modest reduction of CRH-receptor mRNA in the lateral nucleus following the corticosterone treatment (Makino *et al.,* 1995).

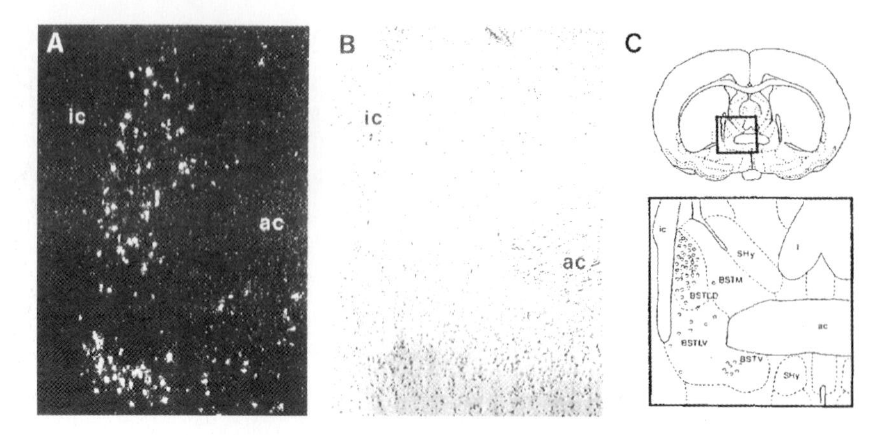

FIGURE 11 Dark-field photomicrograph shows localization of CRH mRNA signals in the bed nucleus of the stria terminals (BNST). Autoradiographic silver grains appear white in A. CRH mRNA is mainly accumulated in the dorsal part of the lateral BNST, and signals are also observed in the ventral part of BNST. Cresyl-violet stained section corresponding to A shows cellular architecture of the region (B). In the schematic representation shown in C, open circles show the distribution of CRH immunoreactive cell bodies. BSTLD, dorsolateral BNST; BSTLV, ventrolateral BNST; BSTM, medial BNST; BSTV, ventral BNST; ac, anterior commisure; f, fornix; ic, internal capsule; SHy, septohypothalamic nucleus (Makino, Gold, and Schulkin, 1994b).

by CRH fibers from both the central nucleus of the amygdala and the bed nucleus of the stria terminalis (Gray, 1990; Swanson *et al.,* 1983) and projects to the central gray and spinal chord (Valentino, Pavcovich, and Hirata, 1995) and appears to be part of the circuitry underlying neuroendocrine regulation of fear (Conti and Foote, 1995).

Thus, there is differential regulation of CRH = producing cells in the brain by glucocorticoid hormones (Swanson and Simmons, 1989; Makino *et al.,* 1994 a,b; Watts and Sanchez-Watts, 1995; see also Imakie *et al.,* 1991; Sawchenko, 1987). Whereas one important role of glucocorticoids is to restrain its own production by the inhibition of hypothalamic-pituitary activation, another is perhaps to sustain the central state of fear (see following discussion).

In another series of studies we tested the hypothesis that pretreatment of glucocorticoids, which raise CRH levels in the amygdala, would facilitate fear-related conditioned freezing responses (Corodimas *et al.,* 1994). Rats were first taught to associate shock with an auditory cue. Conditioned freezing is the typical response. Several days after the last conditioning trial, rats were treated for 5 days with corticosterone or vehicle. When tested again, we found that rats treated with corticosterone, when compared to controls, demonstrated elevated freezing responses when placed in the conditioning chamber. See Figure 13.

We also found some evidence that direct application of corticosterone to the central nucleus of the amygdala facilitates fear-induced freezing behavior to unfamiliar environments (P. Holmes and J. Schulkin, 1994, unpublished observa-

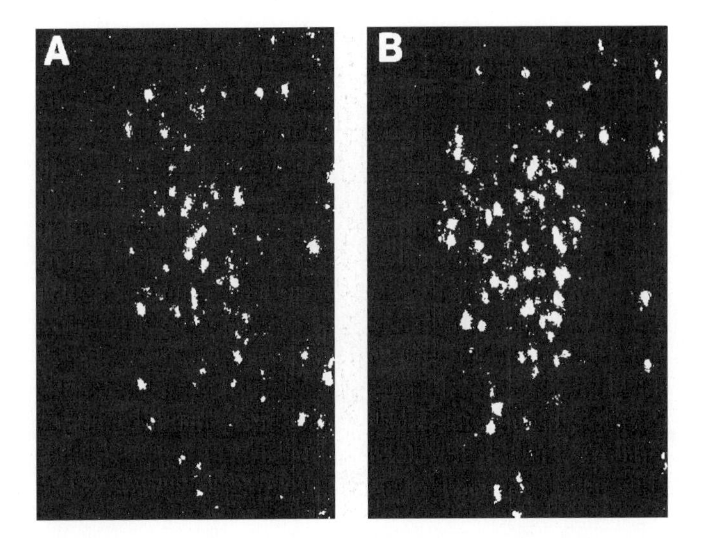

FIGURE 12 Effects of corticosterone (CORT) on CRH mRNA levels in the bed nucleus of the stria terminals (BNST). Dark-field photomicrographs show the autoradiographic distribution of CRH mRNA in the BNST of control (A) and CORT treated rats (B). Autoradiographic silver grains appear white (Makino, Gold, and Schulkin, 1994b).

tions). Thus at least two types of fear responses were facilitated by pretreatment of corticosterone.

Freezing is one expression of fear; startle is another. Therefore, we tested whether pretreatment with glucocorticoids would facilitate CRH-induced startle responses (Lee, Schulkin, and Davis, 1994). Rats were pretreated for several days with corticosterone and then infused centrally with CRH. We chose a dose of CRH which would not normally elicit startle. We found that it did in those rats pretreated with corticosterone, and that the response was specific to corticosterone and not aldosterone (Lee, Y. *et al.,* 1994, unpublished observations).[2]

Finally, high doses of CRH can facilitate seizures that reflect amygdala activation (Weiss *et al,* 1986). Low doses of centrally infused CRH, which by itself does not induce seizures, did so when the rats were pretreated with glucocorticoids (at doses which also did not induce seizures). In other words, instead of reducing the seizures as predicted by glucocorticoid mediated negative feedback of CRH, the glucocorticoids actually potentiated the seizures. That is, the glucocorticoids, by increasing central CRH, lowered the threshold for seizures that is linked to amygdala activation (Rosen *et al.,* 1994). This is a very different outcome than what many in the field would have predicted based on glucocorticoid regulation of CRH gene expression at the level of the PVN.

[2]It has been pointed out that freezing is also linked to states other than fear (e.g., Porges, 1996). The same holds for startle responses. These behavioral measures in order to be meaningful, with regard to fear, have to be functionally linked to states of the brain and other expressions of the animal.

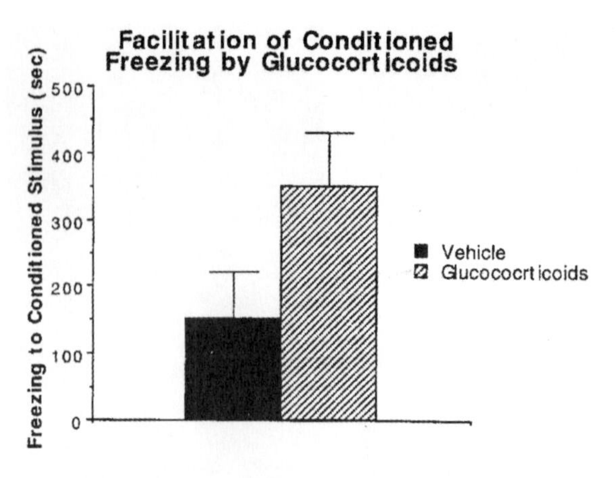

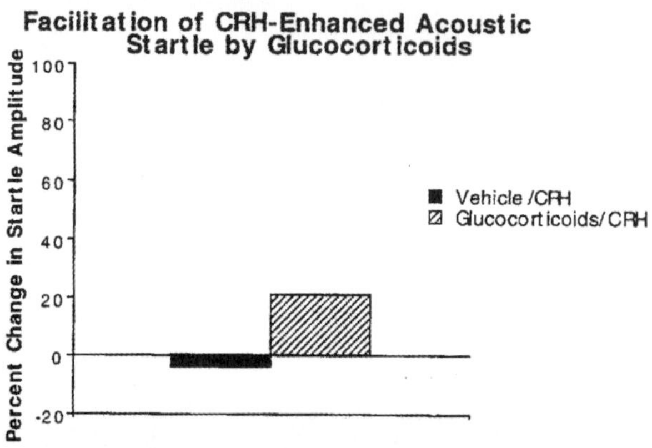

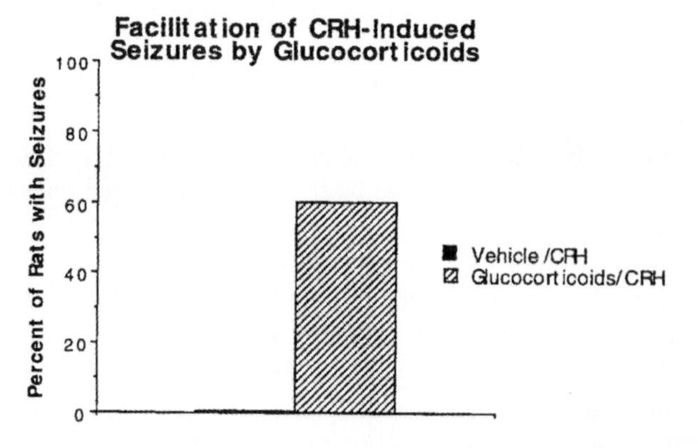

FIGURE 13 The potientating effects on glucocorticoids and CRH on fear-related behaviors. (Top) Glucocorticoids facilitate conditioned-fear-inducing freezing (Corodimas *et al.*, 1994). (Middle) Glucocorticoids facilitate corticotropin-releasing hormone enhanced acoustic startle (Lee *et al.*, 1994). (Bottom) Glucocorticoids facilitate corticotropin-releasing hormone-induced seizures (Rosen *et al.*, 1994).

Thus in addition to restraining physiological events, glucocorticoids facilitate a number of behavioral responses.[3] But the other function of glucocorticoids is to regulate the behavioral responses, to sustain behavior by increasing neuropeptide gene expression, such as CRH. In other words, cortisol influences the expression of central CRH in the brain thereby facilitating behaviors associated with the central motive state of fear. The vigilance associated with sustained fear is linked to elevation of glucocorticoids and central CRH.

Some new observations suggest that CRH mediated effects on long-term anxiety may reflect the bed nucleus of the stria terminalis (Davis, Wasiker, and Lee, 1997). Fear is typically described in the context of a particular object of which one is afraid. By contrast, anxiety is often described as a general state divorced from objects. It has been suggested that the activation of CRH in the amygdala is linked to fear, whereas the activation of CRH in the bed nucleus is linked to anxiety (Davis, Wasiker, and Lee, 1997). In both cases glucocorticoids, in addition to extrahypothalamic CRH, are playing an important role. My colleague Jeff Rosen and I have argued that the aberration of normal fear into pathological anxiety may reflect a transition from normal amygdala function (including extended amygdala) to abnormal function and chronic anxiety and pathology (see also LeDoux, 1996).

Finally consider several other contexts in which CRH is elevated and may be linked to glucocorticoids. Several experimental manipulations, which elevate glucocorticoids in infants, impact CRH expression and the sense of fear in adults. For example, macaques that were raised under conditions in which food availability was uncertain (Coplan *et al.,* 1996) had elevated CRH levels as adults. These same monkeys that had higher CRH than litter mate controls were also more fearful as adults.

Rats maternally deprived as neonates have elevated levels of corticosterone (Levine *et al.,* 1991). As adults they have elevated CRH mRNA in the central nucleus of the amygdala and bed nucleus of the stria terminalis and are more likely than controls to be more fearful (e.g., greater susceptibility to become helpless under uncontrollable adverse conditions, Plotsky, unpublished observations). They also have elevated levels of CRH mRNA in the PVN (Plotsky and Meaney, 1993). And although I have been stressing contexts in which CRH systems in the forebrain can be dissociated, there are also instances in which all three CRH-producing sites are increased. In other words, there are several experimental contexts in which both hypothalamic (PVN; e.g., Raadsheer *et al.,* 1994) and extrahypothalamic CRH (central nucleus of the amygdala and bed nucleus of the stria terminalis) can be elevated at the same time, typically under extreme duress (Koob *et al.,* 1993; Kalin, Takahashi, and Chen, 1994). It is conceivable that in clinical depres-

[3]My remarks about glucocorticoids should not be misconstrued to mean that they are the hormones of fear. They are not. They are the hormones of energy balance, and they help facilitate a number of behaviors (Schulkin, McEwen, and Gold, 1994). The list is diverse and impressive and includes mineralocorticoid-induced sodium ingestion (Ma *et al.,* 1993), central angiotensin-induced water ingestion (Sumners, Gault, and Fregly, 1991), and self-administration of amphetamines (Piazza *et al.,* 1991) or cocaine (Goeders and Guerin, 1996), to name only a few examples.

sion, in which there is extreme fearfulness, there is at the same time a compromised restraint on CRH production in the hypothalamic PVN and a simultaneous sustained increase of CRH in these two extrahypothalamic CRH producing sites.

Finally, CRH in the cerebrospinal fluid has been reported to be elevated in posttraumatic stress disorder patients (Darnell *et al.,* 1994). Interestingly, posttraumatic stress patients have low basal levels of cortisol but hypersecrete cortisol when the HPA axis is normally provoked (Yehuda *et al.,* 1993). This also holds for rape victims and holocaust survivors (Yehuda *et al.,* 1994; 1995). The major point is that they have elevated central CRH.

In each of the conditions cortisol may facilitate the expression of the neuropeptide CRH resulting in the central state of fearfulness. These endocrine events can have occurred years or months before. In this regard this is not dissimilar to organizational effects that occur as a result of gonadal steroid hormones (e.g., Goy and McEwen, 1980; Takahashi, 1995).

VIII. Neuroendocrine Basis of Excessive Shyness and Fear in Children

Approximately 10% to 15% of young children are shy and fearful (Kagan, Reznick, and Snidman, 1988). Shyness appears to be a heritable trait—a temperamental characteristic (Hebb, 1949; Kagan, Reznick, and Snidman, 1988). These traits have been noted in the first few months of life and remain stable over time. Shy children are typically quiet, vigilant, and withdrawn in social contexts. For example, these children are behaviorally inhibited in novel situations and socially wary. In fact, social anxiety (e.g., self-presentation) is a primary feature of shy children and adults. Shy, fearful children remain close to their mothers when presented with novel situations, and they show exaggerated startle responses (Gunnar *et al.,* 1989; Schmidt *et al.,* 1997).

These young children have high levels of cortisol (Kagan, Reznick, and Snidman, 1988; Gunnar *et al.,* 1989; Schmidt *et al.,* 1997). Cortisol remains elevated until at least 7 years of age and is correlated with maternal rating of children's fearfulness and behavioral inhibition in novel contexts (Schmidt *et al.,* in press). See Figure 14.

Interestingly, the endocrine profile of a subset of macaque monkeys resemble these shy children; they are excessively fearful and behaviorally inhibited or shy. Cortisol is elevated in these animals, and they demonstrate longer periods of freezing behavior than other monkeys when placed in novel contexts (e.g., Champoux *et al.,* 1989). Moreover, mothers whose offspring freeze for longer periods also did so (Kalin *et al.,* 1998). Something similar appears to be the case in a subset of human mothers and their children; that is, the mothers of fearful, inhibited, shy children demonstrated greater shyness and greater avoidance behaviors (Richman and Davidson, 1994).

Excessively shy, fearful children are vulnerable to anxiety disorders as adults (Biederman *et al.,* 1990; Windle, 1994). Under these pathological conditions many events are interpreted as fearful when they are not. At the heart of this phenomenon I would speculate should be a hyperactive amygdala (Kagan, 1989; Rosen *et al.,* 1996). One hypothesis is that high cortisol facilitates CRH produc-

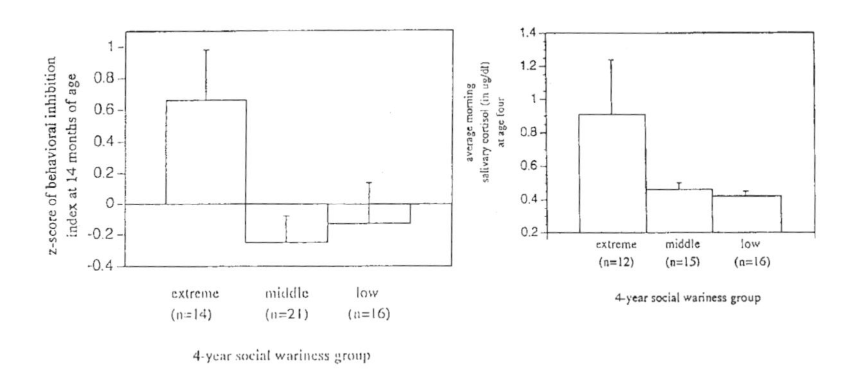

FIGURE 14 Differences among the 3 social groups on behavioral inhibition at 14 months and morning salivary cortisol at age 4 (Schmidt *et al.,* 1997).

tion in the amygdala and/or the extended amygdala (bed nucleus of the stria terminalis) resulting in preparedness to see events as fearful.

In children, one important adaptation to excessive behavioral inhibition or fear is through the sense of attachment (Nachmias *et al.,* 1996). A secure attachment to a parent or care giver is fundamental (Bowlby, 1973) and part of behavioral homeostasis (Hofer, 1994). The desire to secure object constancy may be an exaggerated trait in these children. Interestingly, one adaptation to reduce cortisol levels is linked to their level of attachment to their parents: the greater the degree of estrangement, the greater the degree of circulating cortisol (Nachmias *et al.,* 1996; see also Levine and Wiener, 1988 for a similar phenomenon in monkeys).

Genetic determinants of CRH expression may be fundamentally important in individuals who are vulnerable to excessive fearfulness—excessively shy, fearful children. In this regard, engineered transgenic mice models in which CRH is overproduced in the brain along with greater levels of systemic corticosterone results in greater fear of unfamiliar settings (Stenzel-Poore *et al.,* 1994). See Figure 15. One would predict that excessively shy, fearful children would have greater CRH if one could measure it in their cerebrospinal fluid, as experimenters have demonstrated for fearful, depressed patients and posttraumatic patients.

IX. Conclusion and Some Philosophical Thoughts about the Concept of Fear

The concept of the central motive state of fear, as I use the term, is tied to a view about functionalism. Historically, functionalism was intellectually beholden to biological notions. William James (1890) for example talked about the function of attention, of memory, or of emotions; and he rooted these psychological categories with activity, with survival, with their everyday adaptive use to animals. The study of fear was rooted in biological adaptation, which found expression in the organization of the cen-

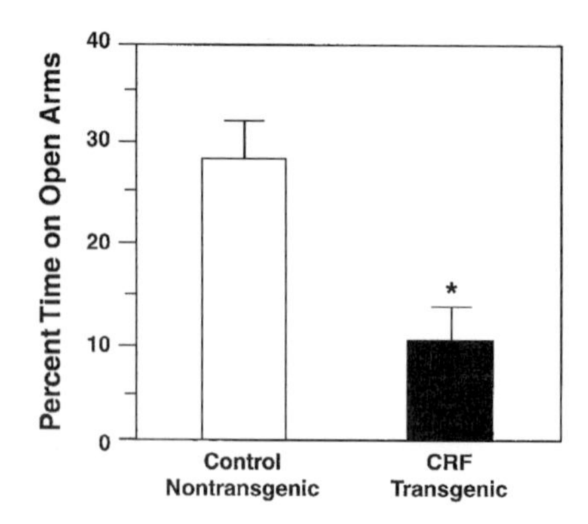

FIGURE 15 Time spent on the open arms of control or corticotropin-releasing factor (CRF) transgenic mice over a 5-min. period on the elevated plus maze (Stenzel-Poore *et al.,* 1994).

tral nervous system. In this regard, the concept of fear was (is) no different from other scientific terms, with its currency anchored to biological and scientific worth.

Note that with the advent of the cognitive sciences there also emerged a second notion of functionalism tied to the way internal mechanisms are wired, along with the idea that perhaps in principle a number of kinds of matter could produce such functional states (e.g., Marr, 1982). In other words, this brand of functionalism grew out of the rise of the cognitive sciences (Simon, 1967; Rey, 1997), and it was linked to appraisal and the evaluation of information.

As I indicated in the introduction, emotional judgements are appraisals. Information is evaluated. These events need not be conscious or even accessible to the animal. In fact, they probably are not. Unconscious evaluations across the spectrum are the rule in the organization of behavior, including fear (e.g., LeDoux, 1996). The mechanisms that render these events possible are unconscious (e.g., Chomsky, 1972; Fodor, 1983; Rozin, 1976). In other words, the mechanisms that make states such as fear possible are not the same as the experience. Central motive states such as fear, therefore, are not synonymous with experience. However, we still want to understand and acknowledge the experiences (Schulkin, 1996; Berridge, 1996).

The cognitive revolution dethroned behaviorism, gave functionalism new life, and evolved quickly in many directions. But the study of the emotions received scant, and at best dismissive, attention (e.g., Fodor, 1983; Churchland and Sejnowski, 1989). Cognition could be characterized and placed in functional terms. But the emotion/motivational systems could not. Perhaps this is now changing (e.g., LeDoux, 1996) because motivational states such as fear are themselves replete with cognition and appraisals of events.

From an evolutionary point of view, motivational states like fear prepare us for action in the face of uncertainty (Dewey, 1894; 1895), which functionally serves to avoid harm. The mechanisms that orchestrate fear are those linked to a readiness to be active, for example, to perceive an event or to generate the requisite motor behavior. All of this is under the rubric of a functional relationship between mental and physical events. Thus two facts about functionalism stand out and have relevance to the concept of the central motive state of fear: the first is the issue of biological adaptation, the second is the internal workings of the central state.

Functionalism never meant, to my mind, that the biological material was irrelevant (as it apparently was understood by a number of investigators [see Rey, 1997]). In fact, the simulation of states like fear or hunger require knowledge of the way brain is organized—what its architecture looks like (e.g., Herrick, 1905; Nauta, 1961) and the way in which information is processed by neural tissue (e.g., LeDoux, 1996; Marr, 1982). But one also needs to discern what steroids and peptides—the "wet material in the brain"—do, and how they do it. In other words, if we are to simulate emotional/motivational states such as fear in matter other than brains, we will need to capture the endocrine brain, which includes understanding gene expression, G proteins, and calcium channels, to name a few molecular events.[4]

The concept of fear, therefore, is a hybrid term: it includes reference to both the brain and psychological events. The key for behavioral neuroscientists is to link them. The concept of central motive states forces one to retain this linkage. My focus in this essay was to link them in a very specific way through steroid/neuropeptide interactions with one particular hypothesis.

Thus in conclusion, there is clearly a set of neural structures that underlie the perception and the behavioral response to frightening events. See Figure 16. I would suggest neuropeptides such as CRH control chemically the sense of fear, which is sustained by elevated cortisol.

Acknowledgments

I thank Kent Berridge and Jeff Rosen for their advice.

References

Abercrombie, H. C., Larson, C. L., Ward, T., Schaefer, S. M., Holden, J. E., Perlman, S. B., Turski, P. A., Krahn, D. D., & Davidson, R. J., Metabolic rate in the amygdala predicts negative affect and depression severity in depressed patients. An FDG-PET study. In press.

Adamec, R. E. (1990). Amygdala kindling and anxiety in the rat. *Neuroreport* **1,** 255–258.

Adolphs, R., Tranel, D., Damasio, H., & Damasio, A. R. (1995). Fear and the human amygdala. *Journal of Neuroscience* **15**(9), 5879–5891.

Aggleton, J. P., & Mishkin, M. (1986). The amygdala: Sensory gateway to the emotions. In R. Plutchik & H. Kellerman (Eds.), *Emotion: Theory, research and experience* (pp. 981–999). Orlando: Academic Press.

[4]There are a number of other investigators who have a similar point of view to the one presented in this essay about the interaction of steroid and neuropeptides in the regulation of central states (e.g., Epstein, 1982; Pfaff, 1982; Herbert, 1993) along with many newer developments about transcription/translational control (Pfaff *et al.,* 1996).

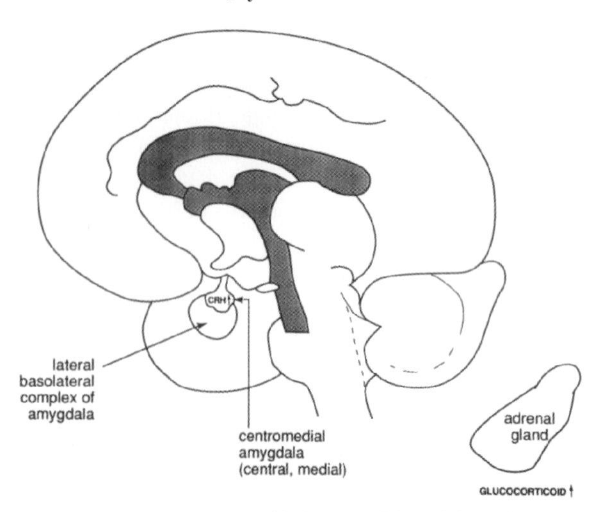

FIGURE 16 A schematic of the human brain with the amygdala and the activation of corticotropin-releasing hormone by glucocorticoids emphasized.

Alheid, G. F., & Heimer, L. (1988). New perspectives in basal forebrain organization and special relevance for neuropsychiatric disorders: The atriatopallidal, amygdaloid and corticipetal components of substantia innominate. *Neuroscience* **22,** 1–39.

Alheid, G. G., deOlmos, J., & Beltramino, C. A. (1996). Amygdala and extended amygdala. In G. Paxinos (Ed.), *The rat nervous system* (2nd ed.). San Diego: Academic Press.

Allman, J., & Brothers, L. (1994). Faces, fear and the amygdala. *Nature* **372,** 613–614.

Amaral, D. G., Price, J. L., Pitkanen, A., & Carmichael, S. T. (1992). Anatomical organization of the primate amygdaloid complex. In J. P. Aggleton (Ed.), *The amygdala: Neurobiological aspects of emotion, memory, and mental dysfunction* (pp. 1–66) New York: Wiley-Liss.

Angrilli, A., Mauri, A., Palomba, D., Flor, H., Birbaumer, N., Sartori, G., & di Paola, F. (1996). Startle reflex and emotion modulation impairment after a right amygdala lesion. *Brain* **119,** 1991–2000.

Aristotle. (384–322 B.C./1941). *The rhetoric, Book 2.* In R. McKeon (Ed.). *The basic works of Aristotle.* New York: Random House.

Armony, J. L., Servan-Schreiber, D., Cohen, J. D., LeDoux, J. E. (1995). An anatomically constrained neural network model of fear conditioning. *Behavioral Neuroscience* **109**(2), 246–257.

Arnold, M. B. (1960). Emotion and anxiety: Sociocultural, biological, and psychological determinants. In A. O. Rorty (Ed.), *Explaining emotions.* Berkeley: University of California Press.

Bard, P. (1939). Central nervous mechanisms for emotional behavior patterns in animals. *Research in Nervous and Mental Disease* **29,** 190–216.

Beach, F. A. (1942). Central nervous mechanisms involved in the reproductive behaviors of vertebrates. *Psychological Bulletin* **26,** 200–225.

Beaulieu, S., DiPaolo, T., Cote, J., Barden, N. (1987). Participation of the central amygdaloid nucleus in the response of adrenacorticotropin secretion to immobilization stress: Opposing roles of neurogenic and dopaminergic systems. *Neuroendocrinology* **45:** 37–46.

Bechara, A., Tranel, D., Damasio, H., Adolphs, R., Rockland, C., & Damasio, A. R. (1993). Double dissociation of conditioning and declarative knowledge relative to the amygdala and hippocampus in humans. *Science* **269,** 1115–1118.

Berridge, K. C. (1996). Food reward: Brain substrates of wanting and liking. *Neuroscience and Biobehavioral Reviews* **20,** 1–25.

Biederman, J., Rosenbaum, J. F., Hirshfield, D. R., Faraone, S. V., Bolduc, E. A., Gersten, M., Meminger, S. R., Kagan, J., Snidman, N., & Reznick, J. S. (1990). Psychiatric correlates of be-

havioral inhibition in young children of parents with and without psychiatric disorders. *Archives of General Psychiatry* **47,** 21–26.
Bindra, D. (1978). How adaptive behavior is produced: A perceptual–motivational alternative to response–reinforcement. *Behavioral and Brain Sciences* **1,** 41–91.
Blanchard, D. C., & Blanchard, R. J. (1972). Innate and conditioned reactions to threat in rats with amygdaloid lesions. *Journal of Comparative Physiology and Psychology* **81,** 281–290.
Booles, R. C., & Fanselow, M. S. (1980). A perceptual–defensive recuperative model of fear and pain. *Brain and Behavioral Sciences* **3,** 291–323.
Bowlby, J. (1973). *Attachment and loss, Vol. II., Separation.* New York: Basic.
Breiter, H. C., Etcoff, N. L., Whalen, P. J., Kennedy, W. A., Rauch, S. L., Buckner, R. L., Strauss, M. M., Hyman, S. E., & Rosen, B. R. (1996). Response and habituation of the human amygdala during visual processing of facial expression. *Neuron* **27,** 875–887.
Brier, A., Albus, M., Pickar, D., Zahn, T. P., & Wolkowitz, O. M. (1987). Controllable and uncontrollable stress in humans: Alterations in mood and neuroendocrine and Psychophysiological function. *American Journal of Psychiatry* **144:** 1419–1425.
Brier, A. (1989). Experimental approaches to human stress research: Assessment of neurobiological mechanisms of stress in volunteers and psychiatric patients. *Biological Psychiatry* **26:** 438–462.
Britton, K. T., Lee, G., Dana, R., Risch, S. C., & Koob, G. F. (1986). Activating and anxiogenic effects of corticotropin-releasing factor are not inhibited by blockade of the pituitary-adrenal axis system with dexamethasone. *Life Science* **39,** 1281–1286.
Cannon, W. B. (1915, 1929). *Bodily changes in pain, hunger, fear and rage.* New York: Harper and Row.
Canteras, N.S., Simerly, R. B., & Swanson, L. W. (1995). Organization of projections from the medial nucleus of the amygdala: A PHAL study in the rat. *Journal of Comparative Neurology* **360:** 213–245.
Carroll, B. J., Curtis, G. C., Davies, B. M., Mendels, J., & Sugarman, A. A. (1976). Urinary free cortisol excretion in depression. *Psychological Medicine* **6,** 43–50.
Champoux, M. D., Coe, C. E., Shankberg, S. M., Kihn, C. M., & Soumi, S. J. (1989). Hormonal effects of early rearing conditions in the infant rhesus monkey. *American Journal of Primatology* **29,** 111–118.
Chomsky, N. (1972). *Language and mind.* New York: Harcourt.
Churchland, P. S., & Sejnowski, T. J. (1989). *The computational brain.* Cambridge: MIT Press.
Conti, L. H., & Foote, S. L. (1995). Effects of pretreatment with corticotropin-releasing factor on the electrophysiological responsivity of the locus coeruleus to subsequent corticotropin-releasing factor challenge. *Neuroscience* **69**(1): 209–219.
Coplan, J. D., Andrews, M. W., Rosenblum, L. A., Owens, M. J., Friedman, S., Gorman, J. M., & Nemeroff, C. B. (1996). Persistent elevations of cerebrospinal fluid concentrations of corticotropin releasing factor in adult nonhuman primates exposed to early life stressors: Implications for the pathophysiology of mood and anxiety disorders. *Proceedings of the National Academy of Science* **93,** 1619–1623.
Corodimas, K. P., LeDoux, J. E., Gold, P., & Schulkin, J. (1994). Corticosterone potentiation of conditioned fear. In R. DeKloet, E. C. Zamita, & P. W. Landfield (Eds.), *Brain corticosteroid receptor.* New York: The New York Academy of Sciences Press.
Cullinan, W. E., Herman, J. P. & Watson, S. J. (1993). Ventral subicular interaction with the hypothalamic paraventricular nucleus: Evidence for a relay in the bed nucleus of the stria terminalis. *Journal of Comparative Neurology.* **332,** 1–20.
Dallman, M. F., Strack, A. M., Akana, S. F., Bradbury, M. J., Hanson, E. S., Scribner, K. A., & Smith, M. (1993). Feast and famine: Critical role of glucocorticoids with insulin in daily energy flow. *Frontiers in Neuroendocrinology* **4,** 303–347.
Damasio, A. R. (1994). *Descartes error: Emotion, reason and the human brain.* New York; Avon.
Darnell, A., Bremner, J. D., Licinio, J., Krystal, J., Nemeroff, C. B., Owens, M., Erdos, J., & Charney, D. S. (1994). Cerebrospiral fluid levels of corticotropin releasing factor in chronic post-traumatic stress disorder. *Neuroscience Abstracts* **20,** 15.
Darwin, C. (1872, 1965). *The expression of emotions in man and animals.* Chicago: University of Chicago Press.

Davidson, R. J. (1994). Asymmetric brain function, affective style and psychopathology: The role of early experience and plasticity. *Development and Psychopathology* **6,** 741–758.
Davis, M., Falls, W. A., Campeau, S., & Kim, M. (1993). Fear-potentiated startle: A neural and pharmacological analysis. *Behavioral Brain Research* **58,** 175–198.
Davis, M., Wasiker, D. L., & Lee, W. L. (1997). Amygdala and bed nucleus of the stria terminalis: Differential roles in fear and anxiety measured with the acoustic startle response. In L. Squire & d. Schacter (Eds.), *Biological and psychological perspectives on memory and memory disorders.* Washington DC: American Psychiatric Press.
DeSouza, E. B. (1987). Corticotropin-releasing factor receptors in the rat central nervous system: Characterization and regional distribution. *Journal of Neuroscience* **7,** 88–100.
Dewey, J. (1894). The theory of emotions I: The theory of the emotions. Emotional attitudes. *Psychological Review* **1,** 553–569.
Dewey, J. (1895). The theory of emotions. II. The significance of emotions. *Psychological Review* **2,** 13–32.
Dickinson, A. (1980). *Contemporary animal learning theory.* Cambridge: Cambridge University Press.
Drevets, W. C., Videen, T. O., Price, J. L., Preskorn, H., Carmichael, S. T., Raichle, M. E. (1992). A functional anatomical study of unipolar depression. *Journal of Neuroscience* **12,** 3628–3641.
Ekman, P., & Davidson, R. J. (1994). *The nature of emotions: Fundamental questions.* New York: Oxford University Press.
Epstein, A. N. (1982). Mineralocorticoids and cerebral angiotensin may act to produce sodium appetite. *Peptides* **3:** 493–494.
Fanselow, M. S. (1994). Neural organization of the defensive behavior system responsible for fear. *Psychonomic Bulletin & Review* **1**(4), 429–438.
Fodor, J. (1981). *Representations. Philosophical essays on the foundations of cognitive science.* Cambridge: MIT Press.
Fodor, J. (1983). *The modularity of mind.* Cambridge: MIT Press.
Fonberg, E. (1972). Control of emotional behavior through the hypothalamus and amygdaloid complex. In D. Hill (Ed.), *Physiology, emotion, and psychosomatic illness* (pp. 131–162) Amsterdam: Elsevier.
Freud, S. (1926, 1960). *A general introduction to psychoanalysis.* New York: Washington Square.
Frijda, N. H. (1986). *The emotions.* Cambridge: Cambridge University Press.
Gallagher, M., & Holland, P. C. (1994). The amygdala complex: multiple roles in associative learning and attention. *Proceedings of the National Academy of Sciences of the USA* **91,** 11771–11776.
Gallistel, C. R. (1980). *The organization of action.* Hillsdale, NJ: Erlbaum Press.
Goeders, N. E., & Guerin, G. F. (1996). Role of corticosterone in intravenous cocaine self-administration in rats. *Neuroendocrinology* **64,** 337–348.
Gold, P. W., Chrousos, G., Kellner, C., Post, R., Avgerinos, P., Schulte, H., & Oldfield, D. (1984). Psychiatric implications of basic and clinic studies with CRF. *American Journal of Psychiatry* **141,** 619–629.
Gold, P. W., Goodwin, F. K., & Chrousos, G. P. (1988). Clinical and biochemical manifestations of depression. [two parts]. *New England Journal of Medicine* **319,** 348–353; 413–420.
Goldstein, K. (1939, 1995). *The organism.* New York: Zone Books.
Goodwin, F. K., & Jamison, K. R. (1990). *Manic-depressive illness.* New York: Oxford University Press.
Gould, E. (1994). The effects of adrenal steroids and excitatory input on neuronal birth and survival. *Annals of the New York Academy of Science,* **743,** 73–92.
Goy, R. W., & McEwen, B. S.(1980). *Sexual differentiation of the brain.* Cambridge: MIT Press.
Gray, J. (1971, 1982). *The psychology of fear and stress.* New York: Cambridge University Press.
Gray, T. S. (1990). The organization and possible function of amygdala CRF. In E. B. DeSouza & C. B. Nemeroff (Eds.), New York: CRC Press. *CRF basic and clinical studies of a neuropeptide.*
Gunnar, M. R., Mangelsdorf, S., Larson, M., & Hartsgaard, L. (1989). Attachment temperament, and adrenocortical activity in infancy; A study of psychoendocrine regulation. *Developmental Psychology* **23,** 355–363.
Hauser, M. D. (1996). *The evolution of communication.* Cambridge: MIT Press.

Hebb, D. O. (1946a). Emotion in man and animal: An analysis of the intuitive processes of recognition. *Psychological Review* **53,** 88–106.

Hebb, D. O. (1946b). On the nature of fear. *Psychological Review* **53,** 258–276.

Hebb, D. O. (1949). *organization of behavior. A neuropsychological theory.*The New York: Wiley.

Herbert, J. (1993). Peptides in the limbic system: Neurochemical codes for co-ordinated adaptive responses to behavioural and physiological demand. *Neurobiology* **41:** 723–791.

Herman, J. P., & Cullinan, W. E. (1997). Neurocircuitry of stress: Central control of the hypothalamo-pituitary-adrenocortical axis. *Trends in Neuroscience* **20**(2), 1–12.

Herrick, C. J. (1905). The central gustatory pathway in the brain of bonyfishes. *Journal of Comparative Neurology* **15,** 375–346.

Hofer, M. A. (1994). Early relationships as regulators of infant physiology and behavior. *Acta Paediatrica Supplement* **397,** 1–9.

Holsboer, F., Muller, O. A., Doerr, H. G., Sippell, W. G., Stalla, G. K., Gerken, A., Steiger, A., Boll, E., & Benkert, O. (1984). ACTH and multisteroid responses to corticotropin-releasing factor in depressive illness: Relationship to multisteroid response after ACTH stimulation and dexamethasone suppression. *Psychoneuroendocrinology* **9,** 147–160.

Honkaniemi, J., Pelto-Huikko, M., Rechardt, L., Isola, J., Lammi, A., Fuxe, K., Gustafsson, J. A., Wikstrom, A. C., Hokfelt, T. (1992). Colocalization of peptide and glucocorticoid receptor immunoreactivities in rat central amygdaloid nucleus. *Neuroendocrinology* **55,** 451.

Imaki, T., Nahan, J.-L., Rivier, C., Sawchenko, P. E., & Vale, W. (1991). Differential regulation of corticotropin-releasing factor mRNA in rat brain regions by glucocorticoids and stress. *Journal of Neuroscience* **11,** 585–599.

James, W. (1884). What is an emotion? *Mind* **9,** 188–205.

James, W. (1890, 1952). *The principles of psychology* (Vols. 1, 2). New York: Dover.

Johnston, J. B. (1923). Further contributions to the study of the evolution of the forebrain. *Journal of Comparative Neurology* **5,** 337–381.

Ju, G., Simerly, R., and Swanson, L. W. (1989). Studies on the cellular architecture of the bed nuclei of the stria terminalis in the rat: 1. Cytoarchitecture. *Journal of Comparative Neurology* **280,** 587–602.

Kagan, J. (1989). *Unstable ideas, temperament, cognition, and self.* Cambridge; Harvard University Press.

Kagan, J., Reznick, J. S., & Snidman, N. (1988). Biological bases of childhood shyness. *Science* **250,** 167–171.

Kagan, J., & Schulkin, J. (1995). On the concepts of fear. *Harvard Review of Psychiatry* **3,** 231–234.

Kainu, T., Honkaniemi, J., Gustafsson, J-A., Rechardt, L., & Markku, P-H. (1993). Co-localization of peptide-like immunoreactivities with glucocorticoid receptor-and fos-like immunoreactivities in the rat parabrachial nucleus. *Brain Research* **615,** 245–251.

Kalin, N. H. (1985). Biological effects of CRF administered to rhesus monkeys. *Federation Proceedings* **44,** 249–253.

Kalin, N. H., Shelton, S. E., Rickman, M., & Davidson, R. J. (1998). Individual differences in freezing and cortisol in infant and mother rhesus monkeys. *Behavioral Neuroscience* **112:** 251–254.

Kalin, N. H., Takahashi, L. K., & Chen, F. L. (1994). Restraint stress increases corticotropin-releasing hormone mRNA content in the amygdala and paraventricular nucleus. *Brain Research* **656,** 182–186.

Kant, G. J., Bauman, R. B., Anderson, S. M., & Moughey, H. (1992). Effects of controllable vs. uncontrollable chronic stress-responsive plasma hormones. *Physiology and Behavior* **51:** 1285–1288.

Kapp, B. S., Frysinger, R. C., Gallagner, M., & Applegate, C. D. (1979). Amygdala central nucleus: Effects of heart rate conditioning in the rabbit. *Physiology and Behavior* **23,** 1109–1117.

Ketter, T. A., Andreason, P. J., George, M. S., Lee, C., Gill, D. S., Parekh, P., Willis, M. W., Herscovitch, P., & Post, R. M. (1996). Anterior paralimbic mediation of procaine-induced emotional and psychosensory experiences. *Archives of General Psychiatry* **53,** 56–69.

Kim, J. J., Rison, R. A., & Fanselow, M. S. (1993). Effects of amygdala, hippocampus, and periaqueductal gray lesions on short- and long-term contextual fear. *Behavioral Neuroscience* **107**(6), 1093–1098.

Kling, A. (1981). Influence of temporal lobe lesions on radio-telemetered electrical activity of amygdala on social stimuli in monkey. In Y. Ben-Ari (Ed.), *The amygdaloid complex* (pp. 271–280) New York: Elsevier.

Konorski, J. (1967). *Integrative activity of the brain.* Chicago: University of Chicago Press.

Koob, G. F., & Bloom, F. E. (1985). Corticotrophin releasing factor and behavior. *Federation Proceedings* **44,** 259–263.

Koob, G. F., Henrichs, S. C., Pich, E. M., Menzahi, F., Baldwin, H., Miczek, K., & Britton, K. T. (1993). The role of corticotropin-releasing factor in behavioral responses to stress. In K. Chadwick, J. Marsh, & K. Ackrill (Eds.), *Corticotropin-releasing factor.* New York: Wiley.

Krettek, J. E., & Price, J. L. (1978). Amygdaloid projections to subcortical structures within the basal forebrain and brainstem in the rat and cat. *Journal of Comparative Neurology* **78,** 225–254.

LaBar, K. S., LeDoux, J. E., Spencer, D. D., & Phelps, E. A. (1995). Impaired fear conditioning following unilateral temporal lobectomy in humans. *Journal of Neuroscience* **15**(10): 6846–6855.

Lance, V. A., & Elsey, R. M. (1986). Stress-induced supression of testosterone secretion in alligators. *Journal of Experimental zoology,* **239,** 241–246.

Lang, P. J. (1995). The emotion probe: Studies of motivation and attention. *American Psychologist* **50,** 372–385.

Lashley, K. S. (1938). An experimental analysis of instinctive behavior. *Psychological Review* **45,** 445–471.

Lazarus, R. S. (1991). *Emotion and adaptation.* New York: Oxford University Press.

LeBar, K. S., LeDoux, J. E., Spencer, D. D., & Phelps, E. A. (1995). Impaired fear conditioning following unilateral temporal lobectomy in humans. *Journal of Neuroscience* **25**(10), 6846–6855.

LeDoux, J. E. (1987). Emotion. *Handbook of Physiology* (pp. 419–459). Bethesda, MD: American Physiological Society.

LeDoux, J. E. (1993). Cognition versus emotion, again this time in the brain: A response to Parrott and Schulkin. *Cognition and emotion,* **7,** 61–64.

LeDoux, J. E. (1995). Emotion: Clues from the brain. *Annual Review of Psychology* **46,** 209–235.

LeDoux, J. E. (1996). *The emotional brain.* New York: Simon & Schuster.

LeDoux, J. E., Cicchetti, P., Xagoraris, A., & Romanski, L. M. (1990). The lateral amygdaloid nucleus: Sensory interfaced of the amygdala in fear conditioning. *Journal of Neuroscience* **10**(4), 1062–1069.

LeDoux, J. E., Iwata, J., Cicceti, P., & Reis, D. J. (1988). Different projections of the central amygdaloid nucleus mediate behavioral correlates of conditioned fear. *Journal of Neuroscience* **8,** 2517–2529.

Lee, Y., Schulkin, J., & Davis, M. (1994). Effect of corticosterone on the enchancement of the acoustic startle reflex by CRH. *Brain Research* **666:** 93–98.

Levine, S., Huchton, D. M., Wiener, S. G., & Rosenfeld, P. (1991). Time course of the effect of maternal deprivation on the HPA axis in the infant rat. *Developmental Psychobiology* **24,** 547–558.

Levine, S., & Wiener, S. G. (1988). Psychoneuroendocrine aspects of mother–infant relationships in non-human primates. *Psychoneuroendocrinology* **13,** 143–154.

Liang, K. C., Melia, K. R., Campeau, S., Falls, W. A., Miserendino, M. J., & Davis, M. (1992). Lesions of the central nucleus of the amygdala, but not the paraventricular nucleus of the hypothalamus, block the excitatory effects of corticotropin-releasing factor on the acoustic startle reflex. *Journal of Neuroscience* **12,** 2313–2320.

Lovenberg, T. W., Chalmers, D. T., Liu, C., & DeSouza, E. B. (1996). CRF receptor subtypes are differentially distributed between the rat central nervous system and peripheral tissues. *Endocrinology* **138,** 4130–4138.

Ma, L. Y., McEwen, B. S., Sakai, R. R., & Schulkin, J. (1993). Glucocorticoids facilitate mineralocorticoid-induced sodium intake in the rat. *Hormones and Behavior* **27,** 240–250.

Mackintosh, N. J. (1975). A theory of attention: Variations in the associability of stimulus and reinforcement. *Psychological Review* **82,** 276–298.

Makino, S., Gold, P. W., & Schulkin, J. (1994a). Corticosterone effects on CRH mRNA in the central nucleus of the amygdala and the paraventricular nucleus of the hypoghalamus. *Brain Research* **640,** 105–112.

Makino, S., Gold, P. W., & Schulkin, S. (1994b). Effects of corticosterone on CRH mRNA and content in the bed nucleus of the stria terminalis; comparison with the effects in the central nucleus of the amygdala and the paraventricular nucleus of the hypothalamus. *Brain Research* **657,** 141–149.

Makino, S., Schulkin, J., Smith, M. A., Pacak, K., Palkovits, M., & Gold, P. W. (1995). Regulation of corticotropin-releasing hormone receptor messenger ribonucleic acid in the rat brain and pituitary by glucocorticoids and stress. *Endocrinology* **136**(10): 4517–4525.

Marler, P., & Hamilton, W. J., III. (1966). *Mechanism of animal behavior.* New York: Wiley.

Marr, D. (1982). *Vision.* San Francisco: W. H. Freeman.

Mason, J. W. (1975). Emotions as reflected in pattern of endocrine integration. in L. Levi (Ed.), *Emotions: Their parameters and measurements* (pp. 143–181). New York: Raven Press.

Mason, J. W., Brady, J. V., & Sidman, M. (1957). Plasma 17-hydroxycorticosteroid levels and conditioned behavior in the rhesus monkey. *Endocrinology* **60**(6), 741–752.

McEwen, B. S., Sakai, R. R., & Spencer, R. L. (1993). Adrenal steroid effects on the brain: Versatile hormones with good and bad effects. In Schulkin, J. (Ed.), (pp. 157–189). *Hormonally induced changes in mind and brain.* San Diego: Academic Press.

Miller N. E. (1957). Experiments on motivation: Studies using psychological, physiological and pharmacological techniques. *Science* **126,** 1271–1278.

Miller, N. E. (1959). Liberalization of basic S-R concepts: Extensions to conflict behavior, motivation and social learning. In S. Koch (Ed.), (Vol. 2), (pp 196–292). *Psychology: A study of a science* New York: McGraw-Hill.

Mineka, S. (1986). Animal models of anxiety-based disorders: Their usefulness and limitations. In A. H. Tuma & J. D. Masur (Eds.), *Anxiety and the anxiety disorders* (pp. 199–244). Hillsdale, NJ: Lawrence Erlbaum.

Mogenson, G. J. (1987). Limbic-motor integration. In A. N. Epstein & J. M. Sprague (Eds.), *Progress in psychobiology and physiological psychology.* New York: Academic Press.

Morgan, C., & Stellar, J. (1950). *Physiological Psychology* (2nd ed.). New York: McGraw-Hill.

Morgan, M. A., & LeDoux, J. E. (1995). Differential contribution of dorsal and ventral medial prefrontal cortex to the acquisition and extinction of conditioned fear in rats. *Behavioral Neuroscience* **109**(4), 681–688.

Morris, J. S., Frith, C. D., Perrett, D. I., Rowland, D., Young, A. W., Calder, A. J., & Dolan, R. J. (1996). A differential neural response in the human amygdala to fearful and happy facial expressions. *Nature* **383,** 812–815.

Mowrer, O. H. (1947). On the dual nature of learning/reinterpretation on conditioning and problem solving. *Harvard Educational Review* **17:** 102–148.

Munck, A., Guyre, P. M., & Holbrook, N. J. (1984). Physiological regulation of glucocorticoids in stress and their regulation to pharmacological actions. *Endocrine Review* **5** 25–44.

Nachmias, M., Gunnar, M., Mangelsdorf, S., Parritz, R. H., & Buss, K. (1996) Behavioral inhibition and stress reactivity: The moderating role of attachment security. *Child Development* **67,** 508–522.

Nauta, W. J. H. (1961). Fiber degeneration following lesions of the amygdloid complex in the monkey. *Journal of Anatomy* **95**(4), 515–531.

Nauta, W. J. H., & Domesick, V. B. (1982). Neural associations of the limbic system. *Neural Basis of Behavior* **10,** 175–206.

Nemeroff, C. B., Widerlov, E., Bissette, G., Walleus, H., Karlsson, I., Eklund, K., Kilts, C. D., Loosen, P. T., & Vale, W. (1984). Elevated concentrations of CSF corticotropin-releasing factor-like immunoreactivity in depressed patients. *Science* **226,** 1342–1343.

Nemeroff, C. B., Krishnan, K. R., Reed, D., Leder, R., Beam, C., & Dunnick, N. R. (1992). Adrenal gland enlargement in major depression: A computed tomographic study. *Archives of General Psychiatry* **49,** 384–387.

Norgren, R. (1976). Taste pathways to hypothalamus and amygdala. *Journal of Comparative Neurology* **166,** 17–30.

Ohman, A. (1993). Fear and anxiety as emotional phenomena: Clinical phenomenoloty, evolutionary perspectives, and information-processing mechanisms. In M. Lewis & J. M. Haviland (Eds.), *Handbook of emotions* (pp. 511–536). New York: Guilford Press.

Owens, M. J., & Nemeroff, C. B. (1991). Physiology and pharmacology of CRF. *Pharmacological Review* **43,** 425–473.

Papez, J. W. (1937). A proposed mechanism of emotion. *Archives of Neurology and Psychiatry* **79,** 217–224.

Parrott, W. G., & Schulkin, J. (1993). Neuropsychology and the cognitive nature of the emotions. *Cognition and Emotion* **1,** 43–59.

Pavcovich, L. A., & Valentino, R. J. (1997). Regulation of a putative neurotransmitter effect of corticotropin-releasing factor: Effects of adrenalectomy. *Journal of Neuroscience* **17**(1), 401–408.

Perrin, M. H., Donaldson, C. J., Chen, R., Lewis, K. A., & Vale, W. W. (1993). Cloning and functional expression of a rat brain CRF receptor. *Endocrinology* **6,** 3058–3061.

Petrovich, G. D., & Swanson, L. W. (1997). Projections from the lateral part of the central amygdalar nucleus to the postulated fear conditioning circuit. *Brain Research.* **763,** 247–254.

Pfaff, D. W. (1980). *Estrogens and brain function: Neural analysis of a hormone-controlled mammalian reproductive behavior.* New York: Springer-Verlag.

Phillips, R. G., & LeDoux, J. E. (1992). Differential contribution of amygdala and hippocampus to cued and contextual fear conditioning. *Behavioral Neuroscience* **106**(2), 274–285.

Piazza, P. V., Maccari, S., Deminiere, J-M., LeMoal, M., Mormede, P., & Simon, H. (1991). Corticosterone levels determine individual vulnerability to amphetamine self-administration. *Neurobiology* **88,** 2088–2092.

Pich, E. M., Koob, G. F., Heilig, M., Menzaghi, F., Vale, W., & Weiss, F. (1993). Corticotropin-releasing factor release from the mediobasal hypothalamus of the rat as measured by microdialysis. *Neuroscience* **55,** 695–707.

Pich, E. M., Lorang, M., Yeganeh, M., Rodriguez, de Fonseca, F., Raber, J., Koob, G. F., & Weiss F. (1995). Increase in extracellular corticotropin-releasing factor-like immunoreactivity levels in the amygdala of awake rats during restraint stress and ethanol withdrawl as measured by microdialysis. *Journal of Neuroscience* **15,** 5439–5447.

Plotsky, P. M. Maternal deprivation effects on CRH expression and behavior in adults rats. Unpublished observations.

Plotsky, P. M., & Meaney, M. J. (1993). Early postnatal experience later hypothalamic corticotropin-releasing factor (CRF) mRNA, median eminence CRF content and stress-induced release in adult rats. *Molecular Brain Research* **18,** 195–200.

Porges, S. W. (1997). Emotion: An evolutionary by-product of the neural regulation of the autonomic nervous system. *Annals of New York Academy of Sciences.* **807,** 62–77.

Raadsheer, F. C., Hoogendijk, W. J. G., Stam, F. C., Tilders, F. J. H., & Swaab, D. F. (1994). Increased numbers of corticotropin-releasing hormone expressing neurons in the hypothalamic paraventiclar nucleus of depressed patients. *Neuroendocrinology* **60,** 436.

Rey, G. (1997). *Contemporary philosophy of mind.* Oxford: Blackwell.

Rescorla, R. A., & Wagner, A. R. (1972). A theory of Pavolian conditioning: Variations in the effectiveness of reinforcement and non-reinforcement. In A. Block & W. Prokasy (Eds.), *Classical conditioning II: Current research and theory.* New York: Appleton-Century Crofts.

Ricardo, J. A., & Koh, E. T. (1978). Anatomical evidence of direct projections from the nucleus of the solitary tract to the hypothalamus, amygdala, and other forebrain structures in the rat. *Brain Research* **153,** 1–26.

Richter, C. P. (1949). Domestication of the Norway rat and its implications for the problem of stress. *Proceedings of the Association for Research of Nervous and Mental Disease* **29,** 19–30.

Richman, M. D., & Davidson, R. J. (1994). Personality and behavior in parents of temperamentally inhibited and unihibited children. *Developmental Psychology* **30,** 346–354.

Rosen, J. B., & Davis, M. (1988). Enhancement of acoustic startle by electrical stimulation of the amygdala. *Behavioral Neuroscience* **102:** 95–102.

Rosen, J. B., Hitchcock, J. M., Sananes, C. B., Miserendino, M. J. D., & Davis, M. (1991). A direct projection from the central nervous of the amygdala to the acoustic startle pathway: Anterograde and retrograde tracing studies. *Behavioral Neuroscience* **105**(6), 817–825.

Rosen, J. B., Hitchcock, J. M., Miserendino, M. J., Falls, W. A., Campeau, S., & Davis, M. (1992). Lesions of the perirhinal cortex, but not of the frontal, medial prefrontal, visual or insular cortex block fear-potentiated startle using a visual conditioned stimulus. *Journal of Neuroscience* **12,** 4624–4633.

Rosen, J. B., Hamerman, E., Sitcoske, M., Glowa, J. R., & Schulkin, J. (1996). Hyperexcitability: Exaggerated fear-potentiated startle produced by partial amygdala kindling. *Behavioral Neuroscience* **110,** 43–50.

Rosen, J. B., Pishevar, S., Weiss, S. B., Smith, M. A., Kling, M., Gold, P., & Schulkin, J. (1994). Glucocorticoid potentiation of CRH-induced seizures. *Neuroscience Letters* **174,** 113–116.

Roy, A., Linnoila, M., Karoum, F., Pickar, D. (1988). Urinary-free cortisol in depressed patients and controls: relationship to urinary indices of noradrenergic function. *Psychological Medicine,* **18,** 93–98.

Rozin, P. (1976). Evolution of intelligence: Access to the cognitive unconscious. In E. Stellar & J. Sprague (Eds.), *Progress in psychobiology and physiological psychology.* New York: Academic Press.

Sabini, J., & Silver, M. (1996). On the possible non-existence of emotions: The passions. *Journal of the Theory of Social Behavior* **26,** 375–398.

Sachar, E. J., Hellman, I., Fukushima, D. K., & Gallagher, T. F. (1970). Cortisol production in depressive illness: A clinical and biochemical clarification. *Archives of General Psychiatry* **23,** 289–298.

Sapolsky, R. M. (1992). *Stress, the aging brain, and the mechanisms of neuron death.* Cambridge: MIT Press.

Sapolsky, R. M., Zola-Morgan, S., & Squire, L. R. (1991). Inhibition of glucocorticoid secretion by the hippocampal formation in the primate. *Journal of Neuroscience* **11**(12), 3695–3704.

Sartre, J. P. (1948). *The emotions.* New York: Philosophical Library.

Sawchenko, P. E. (1987). Evidence for differential regulation of CRF and vasopressin immunoreactivities in parvocellular neurosecretory and autonomic-related projections of the paraventricular nucleus. *Brain Research* **437,** 253–263.

Sawchenko, P. E. (1993). The functional neuroanatomy of corticotropin-releasing factor. In D. J. Chadwick, J. Marsh, & K. Acknill (Eds.), *Corticotropin-releasing factor.* New York: Wiley.

Schmidt, L. A., Fox, N. A., Rubin K. H., Sternberg, E. M., Gold, P. W., Smith, C. C., & Schulkin, J. (1997). Behavioral and neuroendocrine responses in shy children. *Developmental Psychobiology* **30,** 127–140.

Schmidt, L. A., Fox, N. A., Sternberg, E. M., Gold, P. W., Smith, C. C., Schulkin, J. Adrenocortical reactivity and social competence in 7-year-olds. Personality and individual differences. In Press.

Schnierla, T. C. (1959). An evolutionary and developmental of biphasic process underlying approach and withdrawal. In M. R. Jones (Ed.). *The Nebraska Symposium on Motivation.* Lincoln, NE: University of Nebraska Press.

Schulkin, J. (1994). Melancholic depression and the hormones of adversity. *Current Directions in Psychological Sciences* **5,** 41–44.

Schulkin, J. (1996). Pragmatism and the cognitive and neural sciences. *Psychological Report* **78,** 499–506.

Schulkin, J., McEwen, B. S., & Gold, P. W. (1994). Allostasis, amygdala and anticipatory angst. *Neuroscience and Biobehavioral Reviews* **18,** 385–396.

Schwaber, J. S., Kapp, G. S., Higgins, G. A., & Rapp, P. R. (1982). Amygdaloid and basal forebrain directing connections with the nucleus of the solitary tract and the dorsal motor nucleus. *Journal of Neuroscience* **2,** 1424–438.

Selye, H. (1956, 1976). *The stress of life.* New York: McGraw-Hill.

Simon, H. (1967, 1996). *The science of the artificial.* Cambridge: MIT Press.

Smith J. (1977). *The behavior of communicating.* Cambridge: Harvard University Press.

Spinoza, B. (1677, 1955). *On the improvement of the understanding.* New York: Dover Press.

Stellar, E. (1954). The physiology of motivation. *Psychological Review* **61,** 5–22.

Stenzel-Poore, M. P., Heinrichs, S. C., Rivest, S., Koob, G. F., & Vale, W. W. (1994). Overproduction of corticotropin-releasing factor in transgenic mice: A genetic model of anxiogenic behavior. *Journal of Neuroscience* **14,** 2579–2584.

Sumners, C., Gault, T. R., Fregly, M. J. (1991). Potentiation of angiotensin 11-induced drinking by glucocorticoids is a specific glucocorticoid Type 11 receptor (GR)-mediated event. *Brain Research* **552,** 283–290.

Swanson, L. W., & Simmons, D. M. (1989). Differential steroid hormone and neural influences on peptide mRNA levels in CRH cells of the paraventricular nucleus: A hybridization histochemical study in the rat. *Journal of Comparative Neurology* **285,** 413–435.

Swanson, L. W., Sawchenko, P. E., Rivier, J., Vale, W. W. (1983). Organization of ovine corticotropin-releasing factor, immunoreactive cells and fibers in the rat brain: An immunohistochemical study. *Neuroendocrinology* **36,** 165–186.

Swerdlow, N. R., Britton, K. T., & Koob, G. F. (1989). Potentiation of acoustic startle by corticotropin-releasing factor (CRF) and by fear are both reversed by alpha-helical CRF (9-41). *Neuropsychopharmacology* **2,** 285–292.

Swiergiel, A. H., Takahashi, L. K., Rubin, W. W., & Kalin, N. H. (1992). Antagonism of corticotropin-releasing factor receptors in the locus coeruleus attenuates shock-induced freezing in rats. *Brain Research* **487,** 263–68.

Takahashi, L. K. (1994). Organizing action of corticosterone on the development of behavioral inhibition in the preweanling rat. *Developmental Brain Research* **81,** 121–127.

Takahashi, L. K. (1995). Glucocorticoids, the hippocampus, and behavioral inhibition in the preweanling rats. *Journal of Neuroscience* **15**(9), 6023–6034.

Takahashi, L. K., & Rubin, W. W. (1993). Corticosteroid induction of threat-induced behavioral inhibition in preweanling rats. *Behavioral Neuroscience* **107**(5), 860–866.

Tinbergen N. (1951, 1969). *The study of instinct.* Oxford: Oxford University Press.

Turner, B. H., & Herkenham, M. (1992). Thalamoamygdaloid projections in the rat: A test of the amygdala's role in sensory processing. *Journal of Comparative Neurology* **313,** 295–325.

Vale, W., Spiess, J., Rivier, C., & Rivier, J. (1981). Characterization of a 41-residue ovine hypothalamic peptide that stimulates the secretion of corticotropin and d-endorphin. *Science* **213:** 1394–1397.

Valentino, R. J., Page, M. E., van Bockstaele, E., & Aston-Jones, G. (1992). Corticotropin-releasing factor innervation of the locus coeruleus region: Distribution of fibers and sources of input. *Neuroscience* **48,** 689–705.

Valentino, R. J., Pavcovich, L. A., & Hirata, H. (1995). Evidence for corticotropin-releasing hormone projections from Barrington's nucleus t the periaqueductal Gray and dorsal motor nucleus of the vagus in the rat. *Journal of Comparative Neurology* **363,** 402–422.

Watts, A. G. (1996). The impact of physiological stimuli on the expression of corticotropin-releasing hormone (CRH) and other neuropeptide genes. *Frontiers in Neuroendocrinology* **17,** 1–48.

Watts, A. G., & Sanchez-Watts, G. (1995). Region-specific regulation of neuropeptide mRNAs in rat limbic forebrain neurons by aldosterone and corticosterone. *Journal of Physiology* **484,** 721–736.

Weiss S., Post, R. M., Gold, P. W., Chrousos, G., Sullivan, T. L., Walker, D. & Pert, A. (1986). CRF-induced seizures and behavior: Interaction with amygdala kindling. *Brain Research* **372,** 345–351.

Wiersma, A., Baauw, A. D., Bohus, B., & Koolhaas, J. M. (1993). Behavioural activation produced by CRH but not a-helical CRH (CRH-receptor antagonist) when microinfused into the central nucleus of the amygdala under stress-free conditions. *Psychoneuroendocrinology* **20,** 423–32.

Windle, M. (1994). Temperamental inhibition and activation: Hormonal and psychosocial correlates and associated psychiatric disorders. *Personality and Individual Differences.* **17**(1), 61–70.

Yang, X. M., Gorman, A. L., & Dunn, A. J. (1990). The involvement of central noradrenergic systems and CRF in defensive withdrawl behavior in rats. *Journal of Pharmacology and Experimental Therapeutics* **255,** 1064–1070.

Yehuda, R., Boisoneau, D., Lowry, M. T., & Giller, E. L. J. (1995). Dose-response changes in plasma cortisol and lymphocyte glucocorticoid receptors following dexamethasone administration in combat veterans with and without posttraumatic stress disorder. *Archives of General Psychiatry* **52,** 583–593.

Yehuda, R., Kahana, B., Binder-Brynes, K., Southwick, S. M., Mason, J. W., & Giller, E. L. (1995). Low urinary cortisol excretion in Holocaust survivors with posttraumatic stress disorder. *American Journal of Psychiatry* **152,** 982–986.

Yehuda, R., Resnick, H., Kahana, B., & Giller, E. L. (1993). Long lasting hormonal alterations to extreme stress in humans: Normative or maladaptive? *Psychomatic Medicine* **55,** 287–297.

PROGRESS IN PSYCHOBIOLOGY AND PHYSIOLOGICAL PSYCHOLOGY, VOL. 17

Sleep Circuitry, Regulation, and Function: Lessons from c-Fos, Leptin, and Timeless

Priyattam J. Shiromani

Veterans Administration Medical Center and Harvard Medical School
Brockton, Massachusetts 02401

I. Abstract

The phenomenology of the sleep–wake process suggests that molecular events are involved in sleep–wake homeostasis. For example, there is a slow time course of build-up of sleep drive with wakefulness and the dissipation of this drive with sleep. Moreover, cumulative bouts of sleep are necessary to dissipate the sleep drive following sleep loss. This chapter reviews the recent evidence of immediate-early gene activation during sleep–wakefulness. The function of such molecular activation is unknown but may involve sleep–wake homeostasis. To support this hypothesis we draw on examples from other areas of neuroscience. For instance, the identification of leptin has made it possible to understand how feeding might be regulated. Similarly, molecular cascades involving *timeless* and *period* genes underlie circadian timekeeping. Sleep–wakefulness might also be regulated by a coordinated interaction between extracellular and intracellular/molecular events.

II. Introduction

Sleep is defined by REM (rapid eye movement) sleep and non-REM sleep. Non-REM sleep is distinguished by slow waves in the electroencephalogram (EEG), whereas REM sleep is distinguished by low-voltage fast waves that resemble EEG waves seen during wakefulness. Because REM sleep is similar to wakefulness in many ways, this sleep state is often referred to as paradoxical sleep. How the brain orchestrates the shifts in vigilance states is unclear.

In this chapter we will summarize the recent progress that has been made to understand the sleep process especially the newer efforts focused at the molecular level. In other areas of neuroscience, such as feeding and circadian rhythms, impressive gains have been made through the use of modern molecular methods. However, sleep presents several challenges that make it difficult for sleep researchers to utilize many of the experimental approaches taken by investigators in other areas of neuroscience. Perhaps, the biggest obstacle is the lack of an animal model with a clear sleep disturbance (either excessive or too little sleep). There is a canine model of narcolepsy, and important insight into the neurobiology of the human disease is being gained. However, such a model has limited use in molecular protocols, including genetic mapping, because the canine genomic map is

0363–0951/98 $25.00

poorly characterized. The second major hurdle is that multiple brain areas have been implicated in regulating wakefulness, non-REM, and REM sleep, and the neuronal mechanisms that trigger and sustain each behavioral state are poorly understood. As such, it is not clear where in the brain to look for molecular events. Another obstacle is the short, fragmentary nature of sleep. For instance, in rats and mice, the animal models of choice in molecular studies, sleep is characterized by short (8 to 10 min) bouts of non-REM sleep that are often followed by 1.5 to 2 min bouts of REM sleep. This poses a problem in designing experiments that selectively sample individual non-REM sleep and REM sleep bouts. It is unclear how molecular coded events that are relatively slow, such as de novo synthesis of protein, might occur during individual sleep or REM sleep bouts. However, relatively rapid events, such as phosphorylation of proteins (e.g., transcription factors), could be associated with individual sleep–wake bouts, and these in turn could regulate target gene expression. Such coding might constitute one mechanism that resets the need for sleep. Modern molecular techniques permit us to begin the search for these early events. Molecular techniques also provide the level of sensitivity that could make it possible to identify proteins unique to sleep or REM sleep. The function of such molecular coded events might be specialized in specific brain regions (hippocampus for memory consolidation), or globally the molecular cascades could involve cellular restoration.

Given these challenges, the research strategy that we have followed is one that delineates the network underlying sleep–wakefulness and also provides clues about the molecular cascade. Thus, our strategy is to understand the sleep process from the network to the single cell level. Identification of the signal transduction pathways (extracellular and intracellular) will lead to mechanistic studies that will test whether a defect at some point in the chain results in too much sleep (as in narcolepsy) or too little sleep (as in insomnia).

We first summarize the evidence that separate populations of neurons are responsible for generating wakefulness, non-REM, and REM sleep. We then suggest that the regulation of this network is determined by extracellular and intracellular cascades. Identification of such cascades will permit development of animal models with specific gene deletions. Such gene knockout strategies represent a modern approach to understanding behavior in a manner similar to the electrolytic/chemical lesion procedures that have been used to delineate extracellular circuitry.

III. Using c-Fos as a Neuroanatomical Tool to Delineate Sleep–Wake Circuitry

c-*fos* belongs to a family of immediate-early genes that are rapidly and transiently expressed in cells in response to cell signaling (Sheng and Greenberg, 1990). Two of the earliest genes that are expressed belong to the *fos* and *jun* families (Sheng and Greenberg, 1990). The *fos* family includes c-*fos*, *fos* B, and *fra*-1 whereas the jun family includes c-*jun*, *jun* B, and *jun* D (Sheng and Greenberg, 1990). These genes encode proteins that form dimers that bind to a recognition site, referred to as the AP-1 (activator protein 1) binding site, on a target gene and regulates its expression (Franza *et al.*, 1988; Chiu *et al.*, 1988).

fos and *jun* expression has been noted in a variety of conditions (selected references: Aronin *et al.*, 1990; Kornhauser *et al.*, 1990; Shin *et al.*, 1990; Simonato *et al.*, 1991; Sharp *et al.*, 1991). In the suprachiasmatic nucleus (SCN) the phosphorylation of cyclic-AMP response element binding (CREB) protein has been found (Ginty *et al.*, 1993), and the receptor mechanisms leading to transcription are being elucidated (Zhang *et al.*, 1993). Differing thresholds of c-Fos induction have been found (Chan *et al.*, 1993), and receptor stimulation (Luckman *et al.*, 1994) and the temporal pattern of spike activity dictate the level of c-*fos* expression (Fields *et al.*, 1996). Utilizing double-labeling techniques one can identify the chemical nature of the c-Fos positive cells. Retrograde traces can also be used to trace projections thereby delineating connectivity.

It is necessary to be cautious in interpreting the results with c-Fos. Initially c-*fos* expression was considered as a marker of neuronal firing, but depolarization does not always lead to c-*fos* expression (Cole *et al.*, 1989; Hunt *et al.*, 1987; Brennan *et al.*, 1992; Sagar and Sharp, 1993), and cells that are inhibited have not been shown to express c-*fos* (see Sagar and Sharp, 1993). Undoubtedly, some cells that increase firing but do not express c-*fos* will be missed. Also, complex behaviors such as waking and sleep produce widespread changes in neuronal firing. The presence of c-*fos* may not conclusively identify that the common subset of neurons are indeed those that are controlling sleep–wake. For instance, c-Fos will be seen in other regions during waking, for example, the cerebellum (O'Hara *et al.*, 1993). However, removal of the cerebellum does not disturb sleep (Paz *et al.*, 1982).

We now describe the network underlying sleep–wakefulness. Our review will indicate that neurons subserving wakefulness, non-REM sleep, and REM sleep are intermingled not only with each other but also with cells mediating other functions. The question is how to separate the various populations, determine the neurotransmitter identity, and their projections? In this regard, c-Fos has been extremely useful. However, given the caveats associated with c-Fos, it is necessary to use multiple approaches (natural versus pharmacological) to invoke the same behavior (sleep or waking). We anticipate that separate approaches that trigger the same behavior will activate a common subset of neurons and these should display c-Fos. This type of subtractive analysis can be used to identify the common elements underlying the circuit activation during sleep–wakefulness, that is, the network.

A. Brain Regions Responsible for Waking

Brainstem and forebrain mechanisms are implicated in waking. In the brainstem, Moruzzi and Magoun showed that electrical stimulation of the brainstem reticular core produces arousal in an otherwise sedated animal (Moruzzi and Magoun,

1949). Steriade has now shown that the waking-related area is represented by cholinergic neurons of the lateral-dorsal tegmental (LDT) and pedunculo-pontine tegmental (PPT) nuclei (Steriade and McCarley, 1990). These neurons are active when the EEG is desynchronized (during waking and REM sleep). They innervate the thalamus and are responsible for switching the thalamic neurons from an inherent oscillatory pattern (seen during sleep) to a fast pattern (seen during waking and REM sleep). The pontine LDT-PPT cholinergic neurons, however, represent only one portion of the overall influence of the cholinergic system on waking; there is also a forebrain component.

In the forebrain, two neuronal populations are implicated in waking. One population is the cholinergic neurons in the basal forebrain. Szymusiak has recently reviewed the role of the basal forebrain in EEG desynchronization and waking (Szymusiak, 1995). Putative cholinergic basal forebrain neurons have increased activity during waking and project to the thalamus and to cortex (reviewed in Szymusiak, 1995). These "wake-on" basal forebrain cells decrease their firing in response to local warming, which is somnogenic (Alam *et al.,* 1995b). Selective lesion of the basal forebrain cholinergic neurons using the toxin 192 IgG-saporin does not change total amounts of wake (Bassant *et al.,* 1995), indicating redundant wake systems.

The second population is located in the posterior hypothalamus (PH). This population would also include the histaminergic neurons in the tuberomammillary nucleus (TM). PH lesions produce hypersomnolence (Shoham and Teitelbaum, 1982; McGinty, 1969). PH, including TM, neurons have the highest discharge rate during waking and are virtually silent during sleep (Sakai *et al.,* 1990; Szymusiak *et al.,* 1989; Vanni-Mercier *et al.,* 1984). The wake-active PH neurons are hypothesized to interact with anterior hypothalamic-preoptic (BF-POA) cells implicated in sleep. For instance, the wake-active PH neurons are inhibited by warming of the BF-POA (Krilowicz *et al.,* 1994). Conversely, histamine microinjections into BF-POA produce a dose-dependent increase in wake (Lin *et al.,* 1994). Block of histamine synthesis in the POA increases sleep and decreases wake (Lin *et al.,* 1994). Histamine H1 and H2 receptors are postulated to mediate the arousal (Lin *et al.,* 1994).

Norepinephrine (Rajkowski *et al.,* 1994) and serotonin (Jacobs and Azmitia, 1992) have also been implicated in arousal. Dorsal raphe and locus coeruleus (LC) neurons have highest discharge during waking then decrease their firing during non-REM sleep and become virtually silent during REM sleep (summarized in Steriade and McCarley, 1990). Dorsal raphe (Jouvet, 1972) and LC (Jones *et al.,* 1973) lesions produce transient hypersomnia. Recently, applications of beta-receptor agonists at LC targets, such as medial septal nucleus, elicits arousal (Berridge and Foote, 1996). The role of serotonin in arousal is less clear because of the somnogenic influence of tryptophan and 5-HTP (serotonin precursors) (Jouvet, 1972), and the ability of 5-HTP to reverse para-chloro phenylalanine (PCPA) induced insomnia (Denoyer *et al.,* 1989). The differential action of serotonin in modulating the behavioral state might be related to its action on specific serotonin receptors (Bjorvatn *et al.,* 1995).

The conclusion is that forebrain and brainstem regions are involved in waking. The neurotransmitters involved include acetylcholine (basal forebrain and pons), histamine (PH-TM, the only known collection of histaminergic cells in the brain), and norepinephrine (locus coeruleus).

Using c-Fos, we and others have found that when animals are studied following wakefulness (either natural or forced), numerous Fos-ir cells are found in many brain regions (Peng *et al.,* 1995; Tononi *et al.,* 1994; O'Hara *et al.,* 1993; Landis *et al.,* 1992; Grassi-Zucconi *et al.,* 1993; Cirelli *et al.,* 1995a; Pompeiano *et al.,* 1994; Cirelli *et al.,* 1993; Pompeiano *et al.,* 1992). These cells are prominent in areas known to contain cholinergic cells in the basal forebrain and in the pons (LDT/PPT). With sleep, the number of c-Fos-ir cells decreases dramatically (Shiromani *et al.,* 1993; Cirelli *et al.,* 1993; Grassi-Zucconi *et al.,* 1994; Sherin *et al.,* 1995; Basheer *et al.,* 1997). (See Figure 1.) The decline in c-Fos with sleep might be due to the reduction in noradrenergic tone (Cirelli *et al.,* 1996).

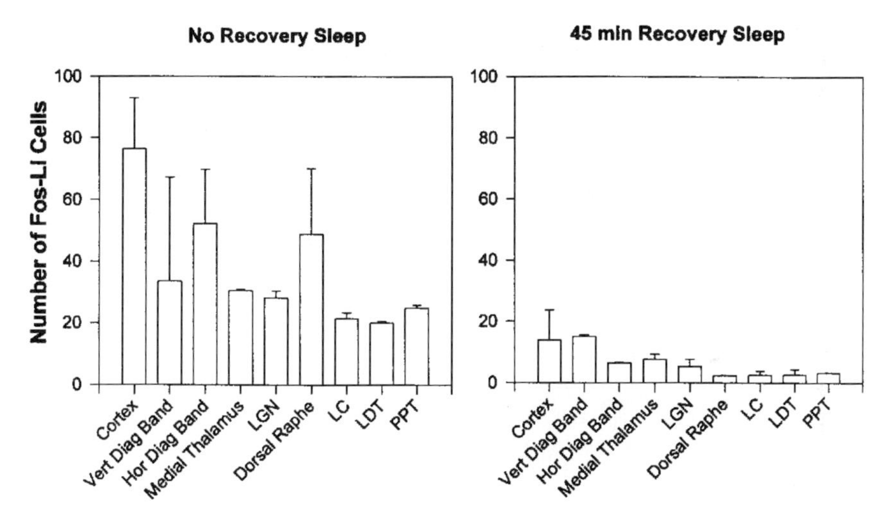

FIGURE 1 The numbers of c-Fos cells decline in response to sleep in many brain regions (Shiromani *et al.,* 1993). Cats were deprived of total sleep for 24 hr by gentle handling. Two cats were killed at the end of the prolonged wakefulness, whereas two other cats were allowed to sleep and then killed. Many brain regions show a decline in number of Fos-positive cells with sleep. Recently, it was demonstrated that the decline might be related to the loss of noradrenergic tone that occurs when locus coeruleus cells reduce firing during sleep (see Cirelli *et al.,* 1996).

We have concentrated on determining c-Fos cells in the basal forebrain where wake-active cells are hypothesized to be cholinergic. Increased release of acetylcholine has been found in the cortex during waking and REM sleep (Szerb, 1967; Jasper and Tessier, 1971). Electrical stimulation of the pontine cholinergic cells also drives cortical acetylcholine release and induces cortical desynchronization (Rasmusson *et al.,* 1994; Rasmusson *et al.,* 1992). Lesions of substantia inominata (SI) and nucleus basalis of Meynert (NBM) produce slow waves in the EEG (Ray and Jackson, 1991; Buzsaki *et al.,* 1988) and also decrease cortical acetylcholine levels (Dekker and Thal, 1993). We have found c-Fos neurons in the NBM. SI, and magnocellular preoptic area (MCPO) following natural waking and following PPT electrical stimulation (see photomicrographs in Figure 2). Note that the distribution of c-Fos in the NBM and SI is similar to the choline acetyltransferase (ChAT) cells. However, we cannot conclude that the c-Fos cells are cholinergic. Nevertheless, because BF cholinergic cells have been implicated in wakefulness these should be c-fos positive. It will be important to determine the distribution of Fos-ir cells in the various BF cell populations.

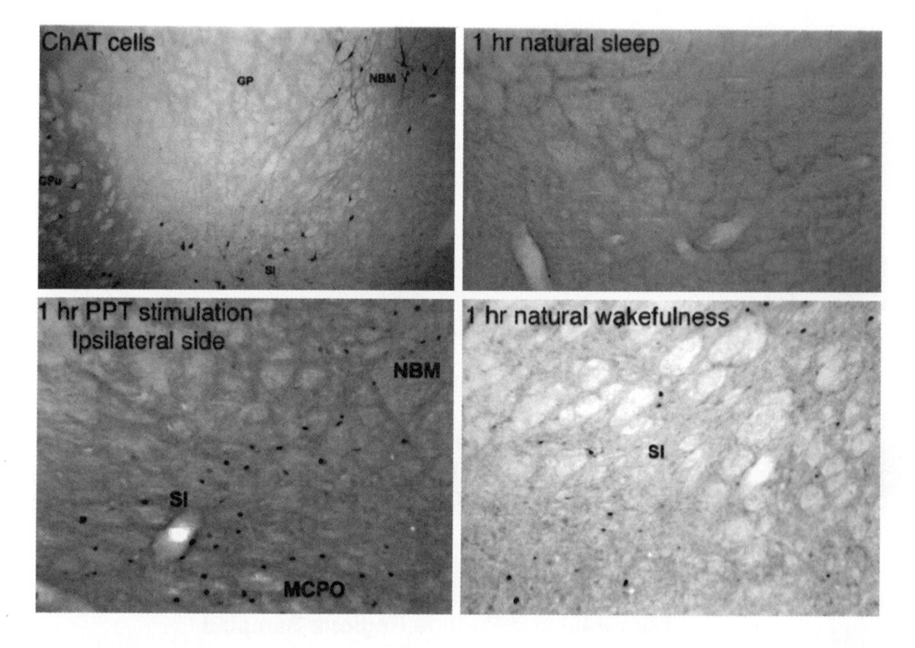

FIGURE 2 Photomicrographs of the rat basal forebrain region. The upper left panel depicts cholinergic somata in the substantia innominata and nucleus basalis. Cells in this region demonstrate c-Fos immunoreactivity during wakefulness or following electrical stimulation (100 Hz, 10 sec on, 10 sec off for 1 hr) of the pedunculopontine tegmental nucleus (PPT). During sleep no c-Fos immunoreactive cells are seen in these cholinergic areas. Abbreviations: ChAT = choline acetyltransferase; cPu = candate putamen; GP = globus pallidus; MCPO = magnocellular preoptic area; NBM = nucleus basalis of Meynert; PPT = pedunculo-pontine tegmental nucleus; SI = substantia inominata.

B. Brain Regions Responsible for Sleep

1. Non-REM Sleep

Converging evidence from lesion (Alam and Mallick, 1990; John *et al.,* 1994; Asala *et al.,* 1990; Shoham *et al.,* 1989; Sallanon *et al.,* 1989; Szymusiak and McGinty, 1986b; Lucas and Sterman, 1975), stimulation (electrical and pharmacological) (Matsumura *et al.,* 1994; Ueno *et al.,* 1982; Denoyer *et al.,* 1989; Mendelson *et al.,* 1989; Sterman and Clemente, 1962a, b) and electrophysiological (detailed in following sections) studies indicates that neurons in the basal forebrain-preoptic area (BF-POA) play an important role in triggering non-REM sleep. For the purposes of brevity we are referring to the BF-POA area as collectively representing the subpopulation of *noncholinergic* sleep-promoting and sleep-active cells. The BF includes the vertical and horizontal limbs of the diagonal band of Broca, the substantia inominata and the magnocellular preoptic area where sleep-active cells have been found (Szymusiak, 1995; Alam *et al.,* 1995a; Szymusiak and McGinty, 1986a). The sleep-active cells are presumed to be GABA and are nested within wake-active (presumably cholinergic) cells (see review Szymusiak, 1995).

The BF-POA is one of only two areas in the brain (the other is in the medulla, Eguchi and Satoh, 1980) where neuronal activity increases during non-REM sleep. In virtually every brain region that has been examined, non-REM sleep is associated with a decline in neuronal discharge. Sleep-active neurons are found in the POA and adjacent basal forebrain in rats, cats, and rabbits (Koyama and Hayaishi, 1994; Findlay and Hayward, 1969; Kaitin, 1984; Szymusiak and McGinty, 1986a). These neurons begin to fire during drowsiness, and peak activity is seen during non-REM sleep. The sleep-active cells comprise about 25% of the recorded cells in the BF-POA and are mixed with wake-active cells, which predominate (Szymusiak, 1995; Koyama and Hayaishi, 1994). Whole-body warming or local warming of the POA induces sleep (summarized in McGinty *et al.,* 1994; Szymusiak *et al.,* 1991; McGinty and Szymusiak, 1990). Such warming also increases activity of many sleep-active neurons and suppresses activity of wake-on neurons in the POA and adjacent basal forebrain (Alam *et al.,* 1995b; McGinty and Szymusiak, 1990). McGinty and Szymusiak (1990) postulate that POA monitors temperature, and sleep (primarily non-REM sleep) might be induced in response to a rise in temperature.

The transmitter identity of the sleep-active neurons is unknown, but there is strong evidence that these could be GABA. For instance, GABA neurons are prevalent in the BF-POA and have projections to the posterior hypothalamus wake-related area (Gritti *et al.,* 1994). Consistent with the descending GABA innervation, Nitz and Siegel (1995) recently reported increased GABA release in PH during natural sleep. Microinjection of muscimol (GABA agonist) into the posterior hypothalamus (PH) induces sleep (Lin *et al.,* 1989). The muscimol induction of sleep is seen in PCPA-treated insomniac cats (Ibid.) and in POA lesioned insomniac cats (Sallanon *et al.,* 1989). BF-POA warming (which is somnogenic) decreases firing of wake-active PH (Krilowicz *et al.,* 1994) and basal forebrain neurons (Alam *et al.,* 1995b).

Because, non-REM active neurons are mixed with wake-REM sleep active cells, c-Fos was used to identify the two populations. Figure 2 depicts the cells in the basal forebrain that show c-Fos following wakefulness, either natural or induced by electrical stimulation of the PPT. During sleep a separate population is seen in the ventral lateral preoptic area (VLPO) (Sherin *et al.,* 1996). Rats were examined following natural periods of sleep. Rats that were mostly asleep (>80%) had many Fos-ir cells in VLPO. Rats that slept very little (<30%) had very few Fos-ir cells in VLPO. To rule out circadian influences on Fos-ir, rats were sleep deprived for 9 to 12 hrs during the day. Some rats were killed immediately at the end of the deprivation, whereas others were allowed varying amounts of recovery sleep. The rats were killed at 4 P.M. (9 hrs deprivation), 7 P.M. (12 hrs deprivation), or 10 P.M. (12 hrs deprivation followed by sleep). We hypothesized that rats killed at 4 P.M. without any sleep should have no Fos-ir cells in VLPO. We further hypothesized that VLPO should demonstrate Fos-ir cells no matter when sleep occurred, including the night cycle. Our hypothesis was confirmed in that Fos-ir cells were seen in VLPO in response to sleep and not due to circadian factors. Rats killed without sleep during all of the time periods showed little or no Fos-ir in VLPO. In contrast, rats given recovery sleep (4 P.M. or 10 P.M.) showed high numbers of Fos-ir cells in VLPO.

Are VLPO cells electrophysiologically active during sleep? Recently, McGinty's group has found a higher percentage of sleep-active cells in the VLPO area (44%) compared to the adjacent preoptic area (Alam *et al.,* 1996). This reinforces our observation of c-Fos cells during natural sleep in VLPO. The VLPO Fos-ir cells project to a group of histaminergic cells in the tuberomammillary nucleus (TM). Such connectivity coupled with the firing profile suggests that VLPO cells could provide an important source of inhibitory influence onto the waking-related TM cells. Neurons in the posterior hypothalamus and the tuberomammillary region, like the LC and dorsal raphe neurons, decrease firing during sleep (Szymusiak *et al.,* 1989; Vanni-Mercier *et al.,* 1984). Our overall hypothesis is that the VLPO cells are part of a group of sleep-active cells in the BF-POA with inhibitory projections to wake-active populations in the histaminergic tuberomammillary nucleus (TM) and the pons (norepinephrine, serotonin, and acetylcholine).

2. REM Sleep

Evidence from lesion, pharmacology, and electrophysiology studies indicates that REM sleep originates from the pons (summarized in Steriade and McCarley, 1990). A number of neurons within the pons (as elsewhere in the brain) markedly increase their firing rates during REM sleep compared to non-REM sleep (for review see Steriade and McCarley, 1990). There are, however, some neurons that progressively increase discharge from waking-to-nonREM-to-REM sleep (Kayama *et al.,*

1992; El Mansari *et al.,* 1989: Steriade *et al.,* 1990; Shiromani *et al.,* 1987a). These cells are called "REM-on" cells. Because of their selective firing during REM sleep, REM-on cells might be considered as possible generators of REMS. The REM-on cells are postulated to represent a subpopulation of the pontine cholinergic cells located in the lateral dorsal tegmental (LDT) and pedunculo-pontine tegmental (PPT) nuclei (Shiromani *et al.,* 1988a).

In order to identify the population of REM-on cells, we used c-Fos immunohistochemistry. In rats and mice, REM sleep lasts for only 1 to 2 min, a period that is too short to detect adequate levels of c-Fos protein. In the cat REM sleep lasts for 5 to 10 min, but this also is too short a time period. Therefore, we chose to induce REM sleep using pharmacological means. Because cholinergic neurons are involved in triggering REM sleep, it is possible to utilize cholinergic agonists to trigger the state. Microinjections of cholinergic agonists into the medial pontine reticular formation have been shown to induce a state that is in every way similar to natural REM sleep (Shiromani *et al.,* 1987b). This state is driven by acetylcholine acting via M2/M3 muscarinic receptors, a finding first shown by us (Velazquez-Moctezuma *et al.,* 1991; Velazquez-Moctezuma *et al.,* 1989) and replicated by others (Imeri *et al.,* 1991; Imeri *et al.,* 1992).

In cats, the cholinergic agonist carbachol induces REM sleep for long periods of time (Shiromani *et al.,* 1987b). We and others have found that during carbachol-induced REM sleep, c-Fos cells are found in the pontine regions implicated in REM sleep (Yamuy *et al.,* 1993; Shiromani *et al.,* 1995b; Shiromani *et al.,* 1992). Because cholinergic REM-on neurons are implicated in triggering REM sleep we hypothesized that some of the c-Fos neurons would also be cholinergic. This was confirmed (Shiromani *et al.,* 1996). We determined using double-labeled immunohistochemistry that about 12% of the pontine LDT-PPT cholinergic cells were also c-Fos positive. Surprisingly, the number of double-labeled cells is quite similar to the percentage of REM-on cells that have been found in this region (Kayama *et al.,* 1992; El Mansari *et al.,* 1989; Steriade *et al.,* 1990; Shiromani *et al.,* 1987). Unfortunately, the double-labeled cells were diffusely represented within the LDT-PPT region, which makes any further detailed analysis quite difficult. On the whole our studies utilizing c-Fos show a convergence with the electrophysiological profile of cells.

IV. Circuit Map of Sleep–Wake Control

Based on the converging lines of evidence presented in the preceding discussion, it is possible to derive a circuit diagram of neuronal centers involved in sleep–wake control. This model builds on features originally articulated by Szymusiak (1995).

The major components of the model involve interaction between GABAergic non-REM sleep-active cells in the basal forebrain-preoptic area (BF-POA), wake-active cells (cholinergic, histaminergic, noradrenergic, and serotonergic), and cholinergic REM-on pontine cells. The VLPO cells constitute a portion of the sleep-active neurons in the BF-POA that are inhibitory to major monoaminergic arousal systems in the basal forebrain, caudal hypothalamus, and pons. The wake-on cells, in turn, would inhibit the sleep-active GABA cells. The inhibitory projections from the sleep-active neurons to the TM and pontine neurons could gate the transition from wake to non-REM sleep and, thereby, also influence REM sleep. Once triggered the pontine REM sleep generator would trigger REM sleep as outlined in the reciprocal-interaction model (Steriade and McCarley, 1990).

This network would be regulated by the SCN. The SCN is the master clock in the brain and one of the cycles it regulates is the circadian expression of sleep–wake rhythms. However, in the absence of the SCN, sleep–wake episodes continue to occur (the total amount of sleep during a 24 hr period is unchanged), but there is no day–night variation in sleep (Ibuka and Kawamura, 1975). In monkeys, however, SCN might regulate wake time (Edgar *et al.,* 1993).

V. Regulation and Function of Sleep: Lessons from Leptin and Timeless

Sleep and wakefulness are regulated by circadian and homeostatic factors (Borbely, 1982). Circadian drive from the SCN regulates sleep timing but there is also a homeostatic component that is more powerful and can override the circadian component (Borbely, 1982). Moreover, in the absence of the circadian oscillator (following SCN lesions) sleep-wakefulness persists although the circadian rhythm is lost. In SCN-lesioned animals sleep time is subject to homeostatic regulation. The relationship between wakefulness and sleep have led many to speculate that an endogenous factor builds during wakefulness and gradually dissipates during sleep. The nature of the homeostatic process is unknown but the phenomenology (build up and dissipation) suggests that it involves intracellular/molecular substrates. Moreover, cumulative bouts of sleep are necessary to dissipate the sleep drive following sleep loss (Levine *et al.,* 1987; Stepanski *et al.,* 1987). To date most of the emphasis has been in identifying the extracellular component. With the development of newer molecular protocols the intracellular cascade is beginning to be understood.

The need for sleep begins with the onset of wakefulness. Many have speculated that an endogenous factor builds during wakefulness and dissipates during sleep. At the electrophysiological level slow wave activity (0.3 to 4 Hz) builds during waking and dissipates across the day cycle, the normal sleep time for rodents (see Borbely, 1982). At the biochemical level numerous endogenous substances have been found to increase with sleep and are thought to serve as sleep factors (see fig. 3). This list includes cytokines (Krueger and Majde, 1994), prostaglandin (Matsumura *et al.,* 1994), and adenosine (Ticho and Radulovacki, 1991). Recently, a new sleep lipid has been found (Cravatt *et al.,* 1995). For all of these agents the mechanism and site of action in the brain needs to be better established. One possible exception is adenosine. Recent evidence shows that adenosine levels are higher in spontaneous

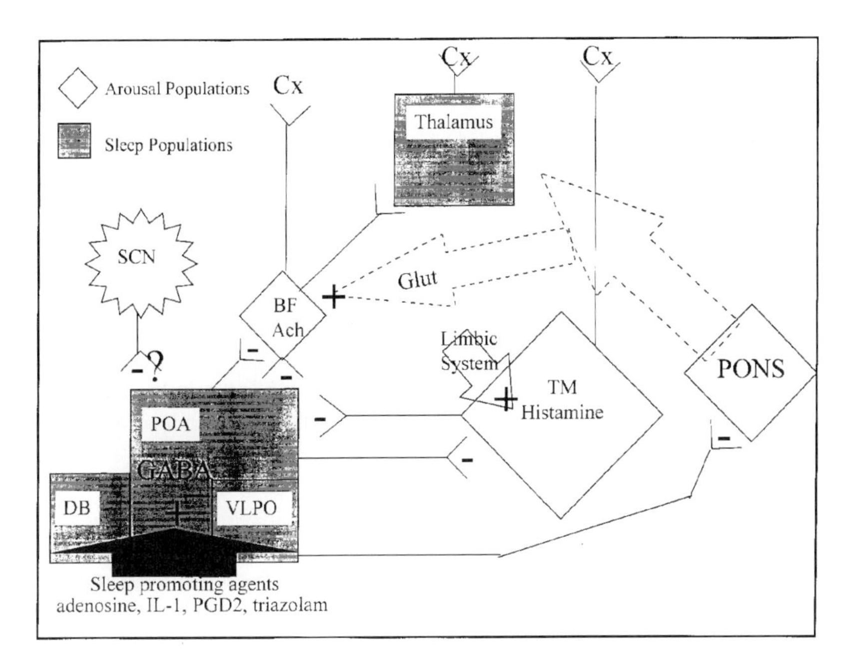

FIGURE 3 Neuronal circuity underlying sleep-wakefulness. Separate populations of cells in the forebrain and pons control behavioral state changes. The arousal populations are represented by cholinergic cells in the basal forebrain and pons, histaminergic cells, and pontine noradrenergic and serotonergic cells. The sleep-active cells are presumably GABAergic and nested within wake–active cells in the preoptic and basal forebrain areas. The pontine REM sleep network is not outlined here but generates REM sleep as hypothesized by the reciprocal-interaction theory (Steriade and McCarley, 1990). Abbreviations: BF Ach: Cholinergic cells in basal forebrain; Cx, Cortex; DB, diagonal band of Broca; Glut, glutamatergic mechanism; POA, preoptic area; SCN, suprachiasmatic nucleus; TM, tuberomammillary nucleus; +, excitatory influence; −, inhibitory influence.

wake than sleep (Porkka-Heiskanen *et al.,* 1996). Adenosine levels also increase with prolonged waking and decrease slowly with sleep. Transport inhibition of adenosine to increase extracellular levels of adenosine levels results in increased sleep. This effect is seen in the basal forebrain cholinergic area but not in the relay nuclei of the thalamus. It is believed that increased adenosine levels during wakefulness promote sleepiness via A1 receptors. Most likely these receptors are localized on basal forebrain and pontine cholinergic neurons where adenosine has been shown to hyperpolarize identified cholinergic cells (Rainnie *et al.,* 1994). Inhibition of the cholinergic wake cells could promote sleep. For the other sleep factors, especially the newer ones such as the sleep lipid, site of action and receptor mechanisms remain to be defined. Recently other investigators have not been able to induce sleep using the sleep lipid (see *The Scientist,* October 28, 1996, p. 13). As for the cytokines, there is a wealth of evidence implicating cytokines in sleep (Krueger and Majde, 1994). However, sites of action are unknown because only systemic or ICV injections have been shown to increase sleep. Similarly, the effect of prostaglandin D2 on sleep needs to be generalized to other mammalian species besides the rat and monkey. Because sleep is ubiquitous, it is reasonable to assume that the process, including the network, are the same across species.

In contrast to data at the extracellular level, very little is known about intracellular sequence of events accompanying sleep. However, intracellular events have been found to dictate the response of the cell to changes at the extracellular level. The intracellular cascade begins with a ligand-receptor interaction followed by changes at the second and third messenger level and eventual activation of target genes. Recent work by a number of investigators has been detailing this cascade as it relates to sleep and wakefulness. Cirelli and Tononi, (1996), have shown that waking is accompanied by increased phosphorylation of protein kinases (as measured by immunohistochemistry). Such phosphorylation could lead to activation of transcription factors, such as pCREB and c-Fos, during waking; and this has been found by us (Basheer et al., 1997) and others (Peng *et al.,* 1995; Tononi *et al.,* 1994; O'Hara *et al.,* 1993; Landis *et al.,* 1992; Grassi-Zucconi *et al.,* 1993; Cirelli *et al.,* 1995; Pompeiano *et al.,* 1994; Cirelli *et al.,* 1993; Pompeiano *et al.,* 1992). The targets of pCREB and c-Fos might be neurotransmitter-synthesizing enzymes such as tyrosine hydroxylase. This has been closely studied in the locus coeruleus. For instance, the noradrenergic locus coeruleus and serotonergic dorsal raphe cells stop firing only during REM sleep. If REM sleep is prevented from occurring, then these neuronal cell groups will continue to discharge. What might be the effects of prolonged periods of REM sleep deprivation on these cells? Because cell firing causes release of neurotransmitters then a prolonged period of discharge of the cell would eventually provoke compensatory mechanisms, such as tyrosine hydroxylase (TH) expression and subsequent TH protein synthesis (Black *et al.,* 1987). We have found increased TH mRNA in LC neurons after 3 and 5 days of REM sleep deprivation (Basheer *et al.,* 1998). We have extended these studies to include the mRNA for the norepinephrine transporter (NET) and have found that NET mRNA increases at the 3-day time point but returns to baseline levels after 5 days of REM deprivation. Other studies have shown changes in neurotransmitters' levels, receptors, and extracellular protein during sleep and following sleep deprivation (Pompeiano and Tononi, 1990; Pompeiano *et al.,* 1990; Tononi *et al.,* 1990; Porkka-Heiskanen *et al.,* 1996; Drucker-Colin *et al.,* 1975). Such changes occurring cumulatively over the course of wakefulness could play an important role in dictating sleep need.

How might extracellular factors regulate sleep need? We believe that this process is conceptually similar to the one that regulates feeding behavior. In the area of feeding it has long been hypothesized that a satiety factor inhibits eating. However, until recently the mechanism underlying feeding was poorly understood. The breakthrough came when it was discovered that obesity in genetically obese mice (ob/ob mice) might be due to the lack of a factor that inhibits eating (reviewed by Miller and Bell, 1996; see also Zhang *et al.,* 1994; Campfield *et al.,*

1995; Halaas *et al.,* 1995; Pelleymounter *et al.,* 1995). The inhibitory factor has been identified to be a 150 amino acid protein made and secreted by adipose tissue and has been named leptin (from the Greek *leptors,* meaning thin). The target of leptin is most likely to be the hypothalamus where it inhibits the activity of neuropeptide Y (NPY) neurons in the arcuate nucleus (Stephens *et al.,* 1995). These neurons in turn project to the paraventricular nucleus. A reduction in the activity of the NPY pathway from the arcuate nucleus to the paraventricular nucleus would decrease feeding (Stanley, 1993). The leptin receptor is a member of the extended cytokine-receptor family (Tartaglia *et al.,* 1995). A defect at the receptor level could halt the signal as has been shown in another strain of mice, the db/db mice, that is also obese (Chen *et al.,* 1996). These mice make plenty of leptin but are obese because of a defect that alters the normal pattern of splicing of leptin-receptor mRNA. Therefore, in the db/db mice the satiety signal (leptin) fails to properly transduce its signal to its receptor, whereas in the ob/ob mice the satiety signal is missing altogether.

In sleep there are few examples in animals or humans where sleep has been found to be changed (either increased or decreased) as a result of inherited alterations in extracellular factors (such as adenosine, any cytokine, prostaglandin, neurotransmitter, etc.). One example is narcolepsy. Human and canine narcolepsy are accompanied by an increased number of muscarinic receptors in the pons (reviewed in Aldrich, 1991). Some of the symptoms of narcolepsy, such as cataplexy and hypnogogic hallucinations, are thought to represent inadvertent intrusions of REM sleep components into waking. As detailed earlier in this chapter, REM sleep is hypothesized to be due to activation of pontine cholinergic mechanisms, specifically via muscarinic receptors. Therefore, in narcolepsy a defect in the cholinergic system, perhaps upregulation of the muscarinic receptor, is a predisposing element of the disease. This possibility is supported by data from a line of rats, the Flinders Sensitive Line (FSL), that show increased number of muscarinic receptors in several brain regions, including the pons. We (Shiromani *et al.,* 1988b) have found that the FSL rats have a selective increase in REM sleep (35%) and that the increase is due to more frequent entry into REM sleep. These findings have been independently confirmed by Benca (personal communication).

We have now extended these studies and found no changes in the number of cholinergic somata in the pons in FSL versus control rats. Moreover, we also find no changes in the mRNA of the various muscarinic receptor subtypes (Greco, Overstreet, and Shiromani, unpublished data). However, the FSL rats have more muscarinic receptors in the pons and in other brain regions. We hypothesize that in the FSL rats and in narcolepsy some change in the mechanism synthesizing the muscarinic receptor (either M2 or M3 receptor subtype) becomes aberrant and manifests itself as increased REM sleep propensity. The nature of the receptor dysfunction in narcolepsy remains to be elucidated. It is also possible that the muscarinic receptor upregulation is due to loss of cholinergic neurons. Siegel has found axonal degeneration in the brains of narcoleptic canines, but it is unclear how this degeneration impacts the cholinergic system (Siegel, paper presented at the 1996 meeting of the Sleep Research Society).

We have cited examples of how changes at the ligand-receptor level might influence sleep. How might molecular events transduce the extracellular signal? This process might be similar to the one found in Drosophila where investigators have shown an exquisite interaction between two proteins, period *(per)* and timeless *(tim)* that dictates clock resetting (Gekakis *et al.,* 1995; Sehgal *et al.,* 1995; Myers *et al.,* 1995). At dawn there is little or no Tim or Per protein, and this permits tim and per genes to be switched on. By midday tim and per transcription is at full strength. At lights off (dusk) Tim and Per protein dimers gain entry into the cell nucleus and begin switching off their respective genes so that Tim and Per protein levels begin to drop. How would light reset the clock? It appears that light destroys Tim protein which then influences transcription and replenishment of the protein (Lee *et al.,* 1996; Myers *et al.,* 1996). On the one hand, an early morning pulse of light would destroy any remaining Tim protein, thereby phase advancing the rhythm by switching on the *tim* and *per* genes earlier. On the other hand, a pulse of light at dusk would destroy some of the accumulating tim protein, thereby delaying the switching off of the *tim* and *per* genes and resulting in a phase-delay.

In sleep, inducible genes could dictate the homeostatic components of sleep. Induction of c-*fos* in specific cell populations during waking and sleep could represent a component of this process. Elimination of c-Fos (as with c-*fos* antisense) in wake populations would attenuate the need for sleep resulting in less sleep. This has been recently shown (Cirelli *et al.,* 1995b). The targets of c-Fos are unknown, but we suggest that the c-Fos cascade represents only one pathway. We have now found that c-*fos* gene knockout mice sleep less (30%) (Shiromani, unpublished observations).

We hypothesize that signal transduction pathways at both extracellular and intracellular levels are necessary in regulating the alternation between wake, non-REM sleep, and REM sleep. Extracellular factors (cytokines, adenosine, neurotransmitters, etc.) would determine the excitability of sleep-active versus wake-active neurons from the single cell to the network level. At the single cell level, receptor and channel openings would influence second messengers (such as cAMP and calcium) resulting in phosphorylation of specific kinases. This would feed back onto the receptor/channel so that membrane excitability could be modulated rapidly (minutes). Activation of transcription factors would transduce the signal to the nucleus and activate specific genes. These genes could encode receptor, neurotransmitter-synthesizing enzymes and enzymes subserving critical cell functions. The activation of these genes would dictate the responsivity of the cell over the long-term (hours to days). Such an alternation between relatively rapid events, such as phosphorylation/dephosphorylation, and long-term events such as target gene expression and protein synthesis could regulate sleep–wake cycles. It is necessary to separate rapid and slow events because in mice and rats, individual bouts of wake and sleep are fairly short (8 to 10 min/cycles), too short a time period for slow cascades such as protein synthesis. Cumulative bouts of sleep are necessary to dissipate the sleep drive. (See Figure 4.)

Sleep should also be a time of replenishment. Recently, Bennington and Heller (1995) have formally proposed that sleep is essential for replenishment of cerebral glycogen stores that are progressively depleted during waking. Their hypothesis specifically focuses on glycogen, but we believe that other evidence of cellular metabolism can also be found to support their general idea. The replenishment might be in key neurotransmitter-synthesizing enzymes. For instance, prolonged periods of cell firing have been associated with increased expression of key neurotransmitter-related enzymes (for review see Black *et al.,* 1987). These changes could represent

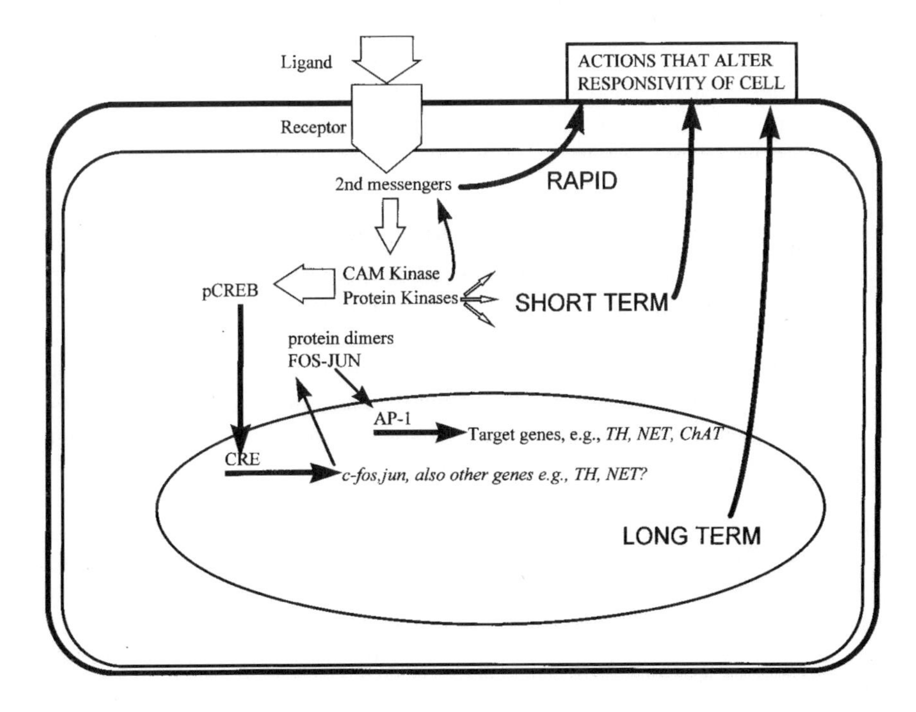

FIGURE 4 Intracellular cascades involved in sleep–wake regulation. Sleep and wakefulness are regulated in large part by homeostatic factors. Rapid and short term intracellular events initiated by ligand-receptor coupling might regulate the transitions between sleep and wakefulness. However, slow cascades such as those involving transcription factors would be necessary to achieve long-term sleep–wake homeostasis. This possibility is supported by findings that show that cumulative bouts of sleep are necessary to dissipate the sleep drive. Abbreviations: AP-1, activator protein-1; ChAT, choline acetyltransferase; CRE, cyclic AMP response elements; NET, norepinephrine transporter; TH, tyrosine hydroxylase.

compensatory mechanisms in response to the neuronal activation. An important question is whether such compensatory changes occur in specific neurons during sleep, a period of neuronal quiescence. Sensitive protocols such as PCR would be needed to identify the changes in mRNA levels. In other procedures, such as subtractive hybridization and differential display, PCR might also be used to identify sleep related genes. Rhyner, Borbely, and Mallet (1990) have used subtractive hybridization procedures and have identified in the cortex several gene transcripts that either increased or decreased expression after 24 hrs of total sleep deprivation. Cirelli, Pompeiano, and Tononi (1996) (see also Pompeiano *et al.,* 1996) used differential display PCR and have identified several mRNA that are differentially expressed in the cortex following spontaneous bouts of waking or sleep. These findings support our hypothesis. It is very likely that the molecular cascades that are initiated are unique to specific cell populations and designed to maintain proper cell function in that network. This is supported by data from Eberwine and colleagues (1992) who have shown that even in homogenous cells specific mRNA is induced. If this is the case then periods of cellular activation and quiescence associated with sleep and waking would serve different functions depending on the cell type. In the hippocampus and cortex gene activation might be involved in cognitive function. Long periods of sleep loss could deplete critical proteins resulting in cognitive loss.

VI. Sleep Changes in Genetic Animal Models

Areas of neuroscience such as feeding, circadian rhythms, epilepsy, memory, and more recently Alzheimer's disease have benefited from the availability of animal models. Chemical mutagenesis can be used to generate animal models, an approach that has been applied successfully in the area of circadian rhythms (Vitaterna *et al.,* 1994). However, in sleep research such an approach is impractical because it would involve screening a large number of animals for signs of aberrant sleep, an extremely labor-intensive process.

The approach that we and others have taken is to outline specific signal transduction pathways that might mediate the sleep process and look at specific points along these pathways. This is a practical and mechanistic approach. Tobler and colleagues (1996) recently demonstrated that mice devoid of the prion protein gene have increased arousals and less non-REM sleep (during the second half of the night cycle) compared to wild-type mice. Moreover, the prion-protein null mice showed increased slow-wave activity following 6 hr total sleep deprivation. These findings have important clinical implications because in the human disease, fatal familial insomnia, a defect in the prion protein gene is suspected to cause the sleep loss (Medori *et al.,* 1992).

We have examined the sleep of mice lacking the calcium-calmodulin dependent protein kinase type II (CaMKII) (Rainnie *et al.,* 1995). CaMKII is a multisubunit serine-threonine protein involved in regulating many cellular functions. These include long-term potentiation (LTP), presynaptic transmitter release, and phosphorylation of tryptophan hydroxylase, the rate-limiting enzyme for serotonin synthesis. The highest concentrations of CaMKII are found in the hippocampus and neocortex. We find no changes in sleep–wake but find a subtle change in the theta frequency only during REM sleep (Thakkar *et al.,* 1997). Hippocampal theta rhythm is a nearly sinusoidal rhythm of about 4 to 10 Hz present in the hippocampus of the animal during exploratory movements of waking and throughout REM sleep. The theta rhythm is present in all the species in which REM occurs and is thought to be involved in learning and memory. These rhythms facilitate the induction of LTP in hippocampal circuits and also facilitate and control the flow of information to the hippocampus or through the hippocampus to the targets. The molecular mechanism of how theta activity induces LTP is not yet clear, though recently Mayford and associates (1995) have proposed that CaMKII is a molecular regulator that regulates the frequency-response function in the theta range at the hippocampal synapses. Our data support this hypothesis.

We envision that the use of gene-knockouts in sleep research will continue. Ultimately, gene deletion studies represent refinements of the knife-cut, transection, and lesion methods that have traditionally been used to understand behavior. Investigators could use excitatory amino acids (kainic and ibotenic), which lesion specific neuronal populations or toxins that target specific neurotransmitter populations, e.g., lesion of basal forebrain cholinergic neurons via IgG192 saporin which binds to the p75 NGF receptor (Bassant *et al.,* 1995). Now the lesion technique has evolved to the point where specific genes can be deleted so that the contribution of that gene to a specific behavior can be better understood. As we gain a better understanding of the molecular cascade underlying sleep, more precise manipulations of the genes (including protein product) in specific cells will be possible. To prepare for such studies it is necessary to understand the sleep process from the single cell to the network level.

Acknowledgments

This work was supported by the DVA Medical Research Service and NIH NS30140.

References

Alam M. N., Szymusiak R., Steininger T. L., & McGinty D. (1996). Sleep-wake activity and thermosensitivity of neurons in the ventral lateral preoptic area of rats. *Society for Neuroscience (Abstracts)* **22,** 27.

Alam, M. N., & Mallick, B. N. (1990). Differential acute influence of medial and lateral preoptic areas on sleep-wakefulness in freely moving rats. *Brain Research* **525,** 242–248.

Alam, M. N., McGinty, D., & Szymusiak, R. (1995a). Neuronal discharge of preoptic/anterior hypothalamic thermosensitive neurons: Relation to NREM sleep. *American Journal of Physiology* **269,** R1240–R1249.

Alam, M. N., Szymusiak, R., & McGinty, D. (1995b). Local preoptic/anterior hypothalamic warming alters spontaneous and evoked neuronal activity in the magno-cellular basal forebrain. *Brain Research* **696,** 221–230.

Aldrich, M. S. (1991). The neurobiology of narcolepsy. *TINS* **14,** 235–239.

Aronin, N., Sagar, S. M., Sharp, F. R., Schwartz, W. J. (1990). Light regulates expression of a fos-related protein in rat suprachiasmatic nuclei. *Proceedings of the National Academy of Science* **87,** 5959–5962.

Asala, S. A., Okano, Y., Honda, K., & Inoue, S. (1990). Effects of medial preoptic area lesions on sleep and wakefulness in unrestrained rats. *Neuroscience Letters* **114,** 300–304.

Basheer, R., Magner, M., McCarley, R. W., & Shiromani, P. (1998). Effects of REM sleep deprivation on TH and NET mRNA levels. *Molecular Brain Research,* In press.

Basheer, R., Sherin, J. E., Saper, C. B., Morgan, J. I., McCarley, R. W., & Shiromani, P. J. (1997). Effects of sleep on wake-induced c-*fos* expression. *The Journal of Neuroscience* **17,** 9746–9750.

Bassant, M. H., Apartis, E., Jazat-Poindessous, F. R., Wiley, R. G., & Lamour, Y. A. (1995). Selective immunolesion of the basal forebrain cholinergic neurons: Effects on hippocampal activity during sleep and wakefulness in the rat. *Neurodegeneration* **4,** 61–70.

Benington, J. H., & Heller, H. C. (1995). Restoration of brain energy metabolism as the function of sleep. *Progress in Neurobiology* **45,** 347–360.

Berridge, C. W., & Foote, S. L. (1996). Enhancement of behavioral and electroencephalographic indices of waking following stimulation of noradrenergic beta-receptors within the medial septal region of the basal forebrain. *The Journal of Neuroscience* **16,** 6999–7009.

Bjorvatn, B., Bjorkum, A. A., Neckelmann, D., & Ursin, R. (1995). Sleep/waking and EEG power spectrum effects of a nonselective serotonin (5-HT) antagonist and a selective 5-HT reuptake inhibitor given alone and in combination. *Sleep* **18,** 451–462.

Black, I. B., Adler, J. A., Dreyfus, C. F., Friedman, W. F., LaGamma, E. F., & Roach, A. H. (1987). Biochemistry of information storage in the nervous system. *Science* **236,** 1263–1268.

Borbely, A. A. (1982). A two process model of sleep regulation. *Human Neurobiology* **1,** 195–204.

Brennan, P. A., Hancock, D., & Keverne, E. B. (1992). The expression of the immediate-early genes c-fos, egr-1 and c-jun in the accessory olfactory bulb during the formation of an olfactory memory in mice. *Neuroscience* **49,** 277–284.

Buzsaki, G., Bickford, R., Ponomareff, G., Thal, L. J., Mandel, R., & Gage, F. H. (1988). Nucleus basalis and thalamic control of neocortical activity in the freely moving rat. *The Journal of Neuroscience* **8,** 4007–4026.

Campfield, L. A., Smith, F. J., Guisez, Y., Devos, R., & Burn, P. (1995). Recombinant mouse OB protein: Evidence for a peripheral signal linking adiposity and central neural networks. *Science* **269,** 546–549.

Chan, R. K. W., Brown, E. R., Ericsson, A., Kovacs, K. J., & Sawchenko, P. E. (1993). A comparison of two immediate-early genes, c-fos and NGFI-B, as markers for functional activation in stress-related neuroendocrine circuitry. *The Journal of Neuroscience* **13,** 5126–5138.

Chen, H., Charlat, O., Tartaglia, L. A., Woolf, E. A., Weng, X., Ellis, S. J., Lakey, N. D., Culpepper, J., Moore, K. J., Breitbart, R. E., Duyk, G. M., Tepper, R. I., & Morgenstern, J. P. Evidence that the diabetes gene encodes the leptin receptor: Identification of a mutation in the leptin receptor gene in db/db mice. *Cell* **84,** 491–495.

Chiu, R., Boyle, W. J., Meek, J., Smeal, T., Hunter, T., & Karin, M. (1988). The c-fos protein interacts with c-jun/AP-1 to stimulate transcription of AP-1 responsive elements. *Cell* **54,** 541–552.

Cirelli, C., Pompeiano, M., & Tononi, G. (1993). Fos-like immunoreactivity in the rat brain in spontaneous wakefulness and sleep. *Archives Italiennes de Biologie* **131,** 327–330.

Cirelli, C., Pompeiano, M., & Tononi, G. (1995a). Sleep deprivation and c-fos expression in the rat brain. *Journal of Sleep Research* **4,** 92–106.

Cirelli, C., Pompeiano, M., & Tononi, G. (1996). Neuronal gene expression in the waking state: A role for the locus coeruleus. *Science* **274,** 1211–1215.

Cirelli, C., Pompeiano, M., Arrighi, P., & Tononi, G. (1995b). Sleep-waking changes after c-fos antisense injections in the medial preoptic area. *Neuroreport* **6,** 801–805.

Cirelli, C., Tononi, G. (1996). Changes in protein phosphorylation patterns in the brain during the sleep-waking cycle. *Society for Neuroscience (Abstracts)* **22,** 688.

Cole, A. J., Saffen, D. W., Baraban, J. M., & Worley, P. F. (1989). Rapid increase of an immediate early gene messenger RNA in hippocampal neurons by synaptic NMDA receptor activation. *Nature* **340,** 474–476.

Cravatt, B. F., Prospero-Garcia, O., Siusdak, G., Gilula, N. B., Henriksen, S. J., Boger, D. L., & Lerner, R. A. (1995). Chemical characterization of a family of brain lipids that induce sleep. *Science* **268,** 1506–1509.

Dekker, A. J., & Thal, L. J. (1993). Independent effects of cholinergic and serotonergic lesions on acetylcholine and serotonin release in the neocortex of the rat. *Neurochemistry Research* **18,** 277–283.

Denoyer, M., Sallanon, M., Kitahama, K., Aubert, C., & Jouvet, M. (1989). Reversibility of parachlorophenylalanine-induced insomnia by intrahypothalamic microinjection of L-5-hydroxytryptophan. *Neuroscience* **28,** 83–94.

Drucker-Colin, R. R., Spanis, C. W., Cotman, C. W., & McGaugh, J. L. (1975). Changes in protein in perfusates of freely moving cats: Relation to behavioral state. *Science* **187,** 963–965.

Eberwine, J., Yeh, H., Miyashiro, K., Cao, Y., Nair, S., Finnell, R., Zettel, M., & Coleman, P. (1992). Analysis of gene expression in single live neurons. *Proceedings of the National Academy of Science USA* **89,** 3010–3014.

Edgar, D. M., Dement, W. C., & Fuller, C. A. (1993). Effect of SCN lesions on sleep in squirrel monkeys: Evidence for opponent processes in sleep-wake regulation. *Journal of Neuroscience* **13,** 1065–1079.

Eguchi, K., & Satoh, T. (1980). Characterization of the neurons in the region of solitary tract nucleus during sleep. *Physiological Behavior* **24,** 99–102.

El Mansari, M., Sakai, K., & Jouvet, M. (1989). Unitary characteristics of presumptive cholinergic tegmental neurons during the sleep-waking cycle in freely moving cats. *Exp Brain Research* **76,** 519–529.

Fields, R. D., Itoh, K., Stevens, B., Eshete, F. (1996). Regulation of c-fos expression by neural impulses: Relation between stimulus pattern, intracellular calcium and CREB phosphorylation. *Society for Neuroscience (Abstracts)* **22.**

Findlay, A. L. R., & Hayward, J. N. (1969). Spontaneous activity of single neurons in the hypothalamus of rabbits during sleep and waking. *Journal of Physiology* **201,** 237–258.

Franza, B. R., Rauscher, F. J., Josephs, S. F., & Curran, T. (1988). The fos complex and fos-related antigens recognize sequence elements that contain AP-1 binding sites. *Science* **239,** 1150–1153.

Gekakis, N., Saez, L., Delahaye-Brown, A., Myers, M. P., Sehgal, A., Young, M. W., & Weitz, C. J. (1995). Isolation of timeless by PER protein interaction: Defective interaction between timeless protein and long-period mutant PER(L). *Science* **270,** 811–815.

Ginty, D. D., Kornhauser, J. M., Thompson, M. A., Bading, H., Mayo, K. E., Takahashi, J. S., & Greenberg, M. E. (1993). Regulation of CREB phosphorylation in the suprachiasmatic nucleus by light and a circadian clock. *Science* **260,** 238–241.

Grassi-Zucconi, G., Giuditta, A., Mandile, P., Chen, S., Vescia, S., & Bentivoglia, M. (1994). c-fos spontaneous expression during wakefulness is reversed during sleep in neuronal subsets of the rat cortex. *Journal of Physiology* **88,** 91–93.

Grassi-Zucconi, G., Menegazzi, M., De Prati, A. C., Bassetti, A., Montagnese, P., Mandile, P., Cosi, C., & Bentivoglio, M. (1993). c-fos mRNA is spontaneously induced in the rat brain during the activity period of the circadian cycle. *European Journal of Neuroscience* **5,** 1071–1078.

Gritti, I., Mainville, L., & Jones, B. E. (1994). Projections of GABAergic and cholinergic basal forebrain and GABAergic preoptic-anterior hypothalamic neurons to the posterior lateral hypothalamus of the rat. *Journal of Comprehensive Neurology* **339,** 251–268.

Halaas, J. L., Gajiwala, K. S., Maffei, M., Cohen, S. L., Chait, B. T., Rabinowitz, D., Lallone, R. L., Burley, S. K., & Friedman, J. M., Weight-reducing effects of the plasma protein encoded by the obese gene. *Science* **269,** 543–546.

Hunt, S. P., Pini, A., & Evans, G. (1980). Induction of c-fos-like protein in spinal cord neurons following sensory stimulation. *Nature* **328,** 632–635.

Ibuka, N., & Kawamura, H. (1975). Loss of circadian rhythm in sleep-wakefulness cycle in the rat by suprachiasmatic nucleus lesions. *Brain Research* **96,** 76–81.

Imeri, L., Bianchi, S., Angeli, P., & Mancia, M. (1991). Differential effects of M2 and M3 muscarinic antagonists on the sleep-wake cycle. *Neuroreport* **2,** 383–385.

Imeri, L., Bianchi, S., Angeli, P., & Mancia, M. (1992). M1 and M3 muscarinic receptors: Specific roles in sleep regulation. *Neuroreport* **3,** 276–278.

Jacobs, B. L., & Azmitia, E. C. (1992). Structure and function of the brain serotonin system. *Physiological Review* **72,** 176–229.

Jasper, H. H., & Tessier, J. (1971). Acetylcholine liberation from cerebral cortex during paradoxical (REM) sleep. *Science* **172,** 601–602.

John, J., Kumar, V. M., Gopinath, G., Ramesh, V., & Mallick, H. (1994). Changes in sleep-wakefulness after kainic acid lesion of the preoptic area in rats. *Japanese Journal of Physiology* **44,** 231–242.

Jones, B. E., Bobillier, P., Pin, C., & Jouvet, M. (1993). The effects of lesions of catecholamine-containing neurons upon monoamine content of the brain and EEG and behavioral waking in the cat. *Brain Research* **58,** 157–168.

Jouvet, M. (1972). The role of monoamines and acetylcholine-containing neurons in the regulation of the sleep-waking cycle. *Ergebnisse Physiology* **64,** 166–308.

Kaitin, K. (1984). Preoptic area unit activity during sleep and wakefulness in the cat. *Exp Neurol* **83,** 347–351.

Kayama, Y., Ohta, M., & Jodo, E. (1990). Firing of "possibly" cholinergic neurons in the rat laterodorsal tegmental nucleus during sleep and wakefulness. *Brain Research* **569,** 210–220.

Kornhauser, J. M., Nelson, D. E., Mayo, K. E., & Takahashi, J. S. (1990). Photic and circadian regulation of c-fos gene expression in the hamster suprachiasmatic nucleus. *Neuron* **5,** 127–134.

Koyama, Y., & Hayaishi, O. (1994). Modulation by prostaglandins of activity of sleep-related neurons in the preoptic/anterior hypothalamic areas in rats. *Brain Research Bulletin* **33,** 367–372.

Krilowicz, B. L., Szymusiak, R., & McGinty, D. (1994). Regulation of posterior lateral hypothalamic arousal related neuronal discharge by preoptic anterior hypothalamic warming. *Brain Research* **668,** 30–38.

Krueger, J. M., & Majde, J. A. (1994). Microbial products and cytokines in sleep and fever regulation. [Review]. *Critical Reviews in Immunology* **14,** 355–379.

Landis, C., Collins, B., Cribbs, L., Smalheiser, N., Sukhatne, V., Bergmann, B., & Rechtschaffen, A. (1992). Expression of EGR-1-like immunoreactivity (EGR-1LI) is altered in specific areas in brain and spinal cord of sleep deprived rats. *Sleep Research* **21,** 321.

Lee, C., Parikh, V., Itsukaichi, T., Bae, K., & Edery, I. (1996). Resetting the Drosophila clock by photic regulation of PER and a PER-TIM complex. *Science* **271,** 1740–1744.

Levine, B., Roehrs, T., Stepanski, E., Zorick, F., & Roth, T. (1987). Fragmenting sleep diminishes its recuperative value. *Sleep* **10,** 590–599.

Lin, J. S., Sakai, K., & Jouvet, M. (1994). Hypothalamo-preoptic histaminergic projections in sleep-wake control in the cat. *European Journal of Neuroscience* **6,** 618–625.

Lin, J. S., Sakai, K., Vanni-Mercier, G., Jouvet, M. (1989). A critical role of the posterior hypothalamus in the mechanisms of wakefulness determined by microinjections of muscimol in freely moving cats. *Brain Research* **479,** 225–240.

Lucas, E. A., & Sterman, M. B. (1975). Effect of a forebrain lesion on the polycyclic sleep-wake cycle and sleep-wake patterns in the cat. *Exp Neurology* **46,** 368–388.

Luckman, S. M., Dyball, R., & Leng, G. (1994). Induction of c-fos expression in hypothalamic magnocellular neurons requires synaptic activation and not simply increased spike activity. *The Journal of Neuroscience* **14,** 4825–4830.

Matsumura, H., Nakajima, T., Osaka, T., Satoh, S., Kawase, K., Kubo, E., Kantha, S., S., Kasahara, K., & Hayaishi, O. (1994). Prostaglandin D2-sensitive, sleep-promoting zone defined in the ventral surface of the rostral basal forebrain. *Proceedings of the National Academy of Science* **91,** 11998–12002.

Mayford, M., Wang, J., Kandel, E. R., & O'Dell, T. J. (1995). CaMKII regulates the frequency-response function of hippocampal synapses for the production of both LTD and LTP. *Cell* **81,** 890–904.

McGinty, D. J. (1969). Somnolence, recovery and hyposomnia following ventromedial diencephalic lesions in the rat. *Electroencephalogram Clinical Neurophysiology* **26,** 70–79.

McGinty, D., & Szymusiak, R. (1990). Keeping cool: A hypothesis about the mechanisms and functions of slow-wave sleep. *Trends in Neurosciences* **13,** 480–487.

McGinty, D., Szymusiak, R., & Thompson, D. (1994). Preoptic/anterior hypothalamic warming increases EEG delta frequency activity within non-rapid eye movement sleep. *Brain Research* **667,** 273–277.

Medori, R., Tritschler, H. J., LeBlanc, A., Villare, F., Manetto, V., Chen, H. Y., Xue, R., Leal, S., Montagna, P., Cortelli P., Tinuper, P., Avoni, P., Mochi, M., Baruzzi, A., Hauw, J. J., Ott, J., Lugaresi, E. Autilio-Gambetti, L., & Gambetti, P. (1992). Fatal familial insomnia, a prion disease with a mutation at Codon 178 of the prion protein gene. *New England Journal of Medicine* **326,** 444–449.

Mendelson, W. B., Martin, J. V., Perlis, M., & Wagner, R. (1989). Enhancement of sleep by microinjection of triazolam into the medial preoptic area. *Neuropsychopharmacology* **2,** 61–66.

Miller, R., J., & Bell, G. I. (1996). JAK/STAT eats the fat. *TINS* **19,** 159–161.

Moruzzi, G., & Magoun, H. W. (1949). Brain stem reticular formation and activation of the EEG. Electroencephalography and clinical. *Neurophysiology* **1,** 455–473.

Myers, M. P., Wager-Smith, K., Rothenfluh-Hilfiker, A., Young, M. W. (1996). Light-induced degradation of TIMELESS and entrainment of the Drosophila circadian clock. *Science* **271,** 1736–1740.

Myers, M. P., Wager-Smith, K., Wesley, C. S., Young, M. W., & Sehgal, A. (1995). Positional cloning and sequence analysis of the drosophila clock gene, timeless. *Science* **270,** 805–808.

Nitz, D. N., & Siegel, J. M. (1995). GABA, glutamate, and glycine release in the posterior hypothalamus across the sleep/wake cycle. *Sleep Research (Abstract)* **24,** 12.

O'Hara, B. F., Young, K., Watson, F., Heller, H. C., & Kilduff, T. S. (1993). Immediate early gene expression in brain during sleep deprivation: Preliminary observations. *Sleep* **16,** 1–7.

Paz, C., Reygadas, E., & Fernandez-Guardiola, A. (1982). Sleep alterations following total cerebellectomy in cats. *Sleep* **3,** 218–226.

Pelleymounter, M. A., Cullen, M. J., Baker, M. B., Hecht, R., Winters, D., Boone, T., & Collins, F. (1995). Effects of the obese gene product on body weight regulation in ob/ob mice. *Science* **269,** 540–543.

Peng, Z. C., Grassi-Zucconi, G., & Bentivoglio, M. (1995). Fos-related protein expression in the midline paraventricular nucleus of the rat thalamus: Basal oscillation and relationship with limbic efferents. *Exp Brain Research* **104,** 21–29.

Pompeiano, M., Cirelli, C., & Tononi, G. (1992). Effects of sleep deprivation on Fos-like immunoreactivity in the rat brain. *Archives Italian Biology* **130,** 325–335.

Pompeiano, M., Cirelli, C., & Tononi, G. (1994). Immediate-early genes in spontaneous wakefulness and sleep: Expression of c-fos and NGFI-A mRNA and protein. *Journal of Sleep Research* **3,** 80–96.

Pompeiano, M., Cirelli, C., & Tononi, G. (1996). Changes in gene expression between wakefulness and sleep revealed by mRNA differential display. *Society for Neuroscience (Abstracts)* **22,** 689.

Pompeiano, M., & Tononi, G. (1990). Changes in pontine muscarinic receptor binding during sleep-waking states in the rat. *Neuroscience Letters* **109,** 347–352.

Pompeiano, M., Tononi, G., & Galbani, P. (1990). Muscarinic receptor binding in forebrain and cerebellum during sleep-waking states in the rat. *Archives Italiennes de Biologie* **128,** 77–79.

Porkka-Heiskanen, T., Strecker, R. E., Bjorkum, A. A., & McCarley, R. W. (1996). Changes in extracellular basal forebrain adenosine levels during spontaneous sleep-waking and prolonged wakefulness. *Society for Neuroscience (Abstracts)* (Vol. 22).

Rainnie, D. G., Grunze, H. C., McCarley, R. W., & Greene, R. W. (1994). Adenosine inhibition of mesopontine cholinergic neurons: Implications for EEG arousal. *Science* **263,** 689–692.

Rainnie, D. G., Shiromani, P. J., Thakkar, M. M., Hearn, E. F., Greene, R. W., & McCarley, R. W. (1995). Changes in sleep and EEG power spectra in CaM-Kinase II knockout mice. *Sleep Research* **24,** 38.

Rajkowski, J., Kubiak, P., & Aston-Jones, G. (1994). Locus coeruleus activity in monkey: Phasic and tonic changes are associated with altered vigilance. *Brain Research Bulletin* **35,** 607–616.

Rasmusson, D. D., Clow, K., Szerb, J. C. (1992). Frequency-dependent increase in cortical acetylcholine release evoked by stimulation of the nucleus basalis magnocellularis in the rat. *Brain Research* **594,** 150–154.

Rasmusson, D. D., Clow, K., & Szerb, J. C. (1994). Modification of neocortical acetylcholine release and electroencephalogram desynchronization due to brainstem stimulation by drugs applied to the basal forebrain. *Neuroscience* **60,** 665–677.

Ray, P., & Jackson, W. (1991). Lesions of nucleus basalis alter ChAT activity and EEG in rat frontal neocortex. *Electroencephalographic Clinical Neurophysiology* **79,** 62–68.

Rhyner, T. A., Borbely, A. A., & Mallet, J. (1990). Molecular cloning of forebrain mRNAs which are modulated by sleep deprivation. *European Journal of Neuroscience* **2,** 1063–1073.

Sagar, S. M., & Sharp, F. R. (1993). Early response genes as markers of neuronal activity and growth factor action. In F. J. Seil (Ed.). *Advances in Neurology* (pp. 273–284). New York: Raven Press.

Sakai, K., ElMansari, M., Lin, J. S., Zhang, G., & Vanni-Mercier, G. (1990). The posterior hypothalamus in the regulation of wakefulness and paradoxical sleep. In (M. Mancia & G. Marini Eds.). *The Diencephalon and Sleep* (pp. 171–198). New York: Raven Press.

Sallanon, M., Denoyer, M., Kitahama, K., Aubert, C., Gay, N., & Jouvet, M. (1989). Long-lasting insomnia induced by preoptic neuron lesions and its transient reversal by muscimol injection into the posterior hypothalamus in the cat. *Neuroscience* **32,** 669–683.

Sehgal, A., Rothenfluh-Hilfiker, A., Hunter-Ensor, M., Chen, Y., Myers, M. P., & Young, M. W. (1995). Rhythmic expression of timeless: A basis for promoting circadian cycles in period gene autoregulation. *Science* **270,** 808–810.

Sharp, F. R., Sagar, S. M., Hicks, K., Lowenstein, D., & Hisanaga, K. (1991). c-fos mRNA, fos, and fos-related antigen induction by hypertonic saline and stress. *Journal of Neuroscience* **11,** 2321–2331.

Sheng, M., & Greenberg, M. E. (1990). The regulation and function of c-fos and other immediate early genes in the nervous system. *Neuron* **4,** 477–485.

Sherin, J. E., Shiromani, P. J., McCarley, R. W., & Saper, C. B. (1996). Activation of ventrolateral preoptic neurons during sleep. *Science* **271,** 216–219.

Sherin, J. E., Shiromani, P. J., Morgan, J., & Saper, C. B. (1995). Recovery sleep in FOS-LacZ mice leads to a rapid decline of the elevated Fos-ir and beta-gal levels which result from forced waking. *Society for Neuroscience (Abstracts).* (Vol. 21, p. 1679).

Shin, C., McNamara, J. O., Morgan, J. I., Curran, T., & Cohen, D. R. (1990). Induction of c-fos mRNA expression by afterdischarge in the hippocampus of naive and kindled rats. *Journal of Neurochemistry* **55,** 1050–1055.

Shiromani, P. J., Armstrong, D. M., Berkowitz, A., Jeste, D. V., & Gillin, J. C. (1988a). Distribution of choline acetyltransferase immunoreactive somata in the feline brainstem: Implications for REM sleep generation. *Sleep* **11,** 1–16.

Shiromani, P., Armstrong, D. M., Bruce, G., Hersh, L. B., Groves, P. M., & Gillin, J. C. (1987a). Relation of choline acetyltransferase immunoreactive neurons with cells which increase discharge during REM sleep. *Brain Research Bulletin* **18,** 447–455.

Shiromani, P., Gillin, J. C., Henriksen, S. J. (1987b). Acetylcholine and the regulation of REM sleep: Basic mechanisms and clinical implication for affective illness and narcolepsy. *Annual Review of Pharmacology and Toxicology* **27,** 137–156.

Shiromani, P. J., Kilduff, T. S., Bloom, F. E., & McCarley, R. W. (1992). Cholinergically induced REM sleep triggers Fos-like immunoreactivity in dorsolateral pontine regions associated with REM sleep. *Brain Research* **580,** 351–357.

Shiromani, P. J., Magner, M., Winston, S., & Charness, M. E. (1995a). Time course of phosphorylated CREB and Fos-like immunoreactivity in the hypothalamic supraoptic nucleus after salt loading. *Molecular Brain Research* **29,** 163–171.

Shiromani, P. J., Mallik, M., Winston, S., McCarley, R. W. (1995b). Time course of Fos-LI associated with carbachol induced REM sleep. *The Journal of Neuroscience* **15,** 3500–3508.

Shiromani, P. J., Overstreet, D., Levy, D., Goodrich, C., Campbell, S. S., & Gillin, J. C. (1988b). Increased REM sleep in rats genetically bred for cholinergic hyperactivity. *Neuropsychopharmacology* **1,** 127–133.

Shiromani, P. J., Winston, S., & McCarley, R. W. (1996). Pontine cholinergic neurons show Fos-like immunoreactivity associated with cholinergically-induced REM sleep. *Molecular Brain Research* **38,** 77–84.

Shiromani, P., Winston, S., Mallik, M., & McCarley, R. W. (1993). Rapid decline in Fos-LI in association with recovery sleep that follows total sleep deprivation. *Society for Neuroscience (Abstracts)* (Vol. 19, Suppl. Pts. 1 and 2).

Shoham, S., Blatteis, C. M., & Krueger, J. M. (1989). Effects of preoptic area lesions on muramyl dipeptide-induced sleep. *Brain Research* **476,** 396–399.

Shoham, S., & Teitelbaum, P. (1982). Subcortical waking and sleep during lateral hypothalamic "somnolence" in rats. *Physiological Behavior* **28,** 323–333.

Simonato, M., Hosford, D. A., Labiner, D. M., Shin, C., Mansbach, H. H., & McNamara, J. O. (1991). Differential expression of immediate early genes in the hippocampus in the kindling model of epilepsy. *Molecular Brain Research* **11,** 115–124.

Stanley, B. G. (1993). The biology of neuropeptide. Y. In W. F. Colmers and C. Wahlestedt (Eds.). *The Biology of Neuropeptide Y and Related Peptides* (pp. 457–509). New York: Humana Press.

Stepanski, E., Lamphere, J., Roehrs, T., Zorick, F., & Roth, T. (1987). Experimental sleep fragmentation in normal subjects. *International Journal of Neuroscience* **33,** 207–214.

Stephens, T. W., Basinski, M., Bristow, P. K., Bue-Valleskey, J. M., Burgett, S. G., Craft, L., Hale, J., Hoffmann, J., Hsiung, H. M., & Kriauciunas, A., et al. (1995). The role of neuropeptide Y in the antiobesity action of the obese gene product. *Nature* **377,** 530–532.

Steriade, M., Datta, S., & Oakson, G. (1990). Neuronal activities in brain-stem cholinergic nuclei related to tonic activation processes in thalamocortical systems. *Journal of Neuroscience* **10,** 2541–2559.

Steriade, M., & McCarley, R. W. (1990). *Brainstem control of wakefulness and sleep.* New York: Plenum Press.

Sterman, M. B., & Clemente, C. (1962a). Forebrain inhibitory mechanisms: Sleep patterns induced by basal forebrain stimulation in the behaving cat. *Experimental Neurology* **6,** 103–117.

Sterman M. B., Clemente, C. (1962b). Forebrain inhibitory mechanisms: cortical synchronization induced by basal forebrain stimulation. *Experimental Neurology* **6,** 9–102.

Szerb, J. C. (1967). Cortical acetylcholine release and electroencephalographic arousal. *Journal of Physiology (London),* **192,** 329–345.

Szymusiak, R. (1995). Magnocellular nuclei of the basal forebrain: Substrates of sleep and arousal regulation. *Sleep* **18,** 478–500.

Szymusiak, R., Danowski, J., & McGinty, D. (1991). Exposure to heat restores sleep in cats with preoptic/anterior hypothalamic cell loss. *Brain Research* **541,** 134–138.

Szymusiak, R., Iriye, T., & McGinty, D. (1989). Sleep-waking discharge of neurons in the posterior lateral hypothalamic area of cats. *Brain Research Bulletin* **23,** 111–120.

Szymusiak, R., & McGinty, D. (1986a). Sleep-related neuronal discharge in the basal forebrain of cats. *Brain Research* **370,** 82–92.

Szymusiak, R., & McGinty, D. (1986b). Sleep suppression following kainic acid-induced lesions of the basal forebrain. *Exp Neurology* **94,** 598–614.

Tartaglia, L. A., Dembski, M., Weng, X., Deng, N., Culpepper, J., Devos, R., Richards, G. J., Campfield, L. A., Clark, F. T., Deeds, J., et al. (1995). Identification and expression cloning of a leptin receptor, OB-R. *Cell* **83,** 1263–1271.

Thakker, M., Rainnie, D. G., Hearn, E. F., Greene, R. W., McCarley, R. W., & Shiromani, P. J. (1997). Abnormal theta activity during REM sleep in α-calcium-calmodulin kinase II knockout mice. *Sleep Research* **26,** 53.

Ticho, S. R., & Radulovacki, M. (1991). Role of adenosine in sleep and temperature regulation in the preoptic area of rats. *Pharmacology and Biochemical Behavior* **40,** 33–40.

Tobler, I., Gaus, S. E., Deboer, T., Achermann, P., Fischer, M., Rhlicke, T., Moser, M., Oesch, B., McBride, P. A., & Manson, J. C. (1996). Altered circadian activity rhythms and sleep in mice devoid of prion protein. *Nature* **380,** 639–642.

Tononi, G., Pompeiano, M., & Cirelli, C. (1994). The locus coeruleus and immediate-early genes in spontaneous and forced wakefulness. *Brain Research Bulletin* **35,** 589–596.

Tononi, G., Pompeiano, M., Ronca-Testoni, S. (1990). Noradrenergic receptor binding during sleep-waking states in the rat. *Archives Italiennes de Biologie* **128,** 67–76.

Ueno, R., Ishikawa, Y., Nakayama, T., & Hayaishi, O. (1982). Prostaglandin D2 induces sleep when microinjected into the preoptic area of the conscious rat. *Biochem Biophys Res Comm* **109,** 576–582.

Vanni-Mercier, G., Sakai, K., & Jouvet, M. (1986). Neurones specifiques de l'eveil dans l'hypothalamus posterieur. *CR Academy of Science* **298,** 195–200.

Velazquez-Moctezuma, J., Gillin, J. C., & Shiromani, P. J. (1989). Effect of specific M1,M2 muscarinic receptor agonists on REM sleep generation. *Brain Research* **503,** 128–131.

Velazquez-Moctezuma, J., Shalauta, M., Gillin, J. C., Shiromani, P. J. (1991). Cholinergic antagonists and REM sleep generation. *Brain Research* **543,** 175–179.

Vitaterna, M. H., King, D. P., Chang, A. M., Kornhuaser, J. M., Lowrey, P. L., McDonald, J. D., Dove, W. F., Pinto, L. H., Turek, F. W., & Takahashi, J. S. (1994). Mutagenesis and mapping of a mouse gene, Clock, essential for circadian behavior. *Science* **264,** 719–725.

Zhang, Y., Proenca, R., Maffei, M., Barone, M., Leopold, L., & Friedman, J. M. (1994). Positional cloning of the mouse obese gene and its human homologue. *Nature* **372**(6505), 425–432.

Zhang, Y., Lee, P. C., Kirby, J. D., Takahashi, J. S., & Turek, F. W. (1993). A cholinergic antagonist, mecamylamine, blocks light-induced fos immunoreactivity in specific regions of the hamster suprachiasmatic nucleus. *Brain Research* **615,** 107–112.

The Locus Coeruleus–Noradrenergic System as an Integrator of Stress Responses

Rita J. Valentino*, Andre L. Curtis*, Michelle E. Page†, Luis A. Pavcovich*, Sandra M. Lechner††, and Elisabeth Van Bockstaele§

Department of Psychiatry, Allegheny University*
Philadelphia, Pennsylvania 19129

Department Psychiatry†, University of Pennsylvania
Philadelphia, Pennsylvania 19104

Allelix Neuroscience††
South Plainfield, New Jersey 07080

Department of Pathology and Anatomy§, Thomas Jefferson University
Philadelphia, Pennsylvania 19107

I. Introduction

The stress response is a classic example of how the brain exerts a coordinated influence over multiple physiological systems in response to a single stimulus to maintain homeostasis. This response is characterized by endocrine, neuronal, and immunological events that are highly coordinated and interactive, producing a powerful net result that is both adaptive, and under some circumstances, pathological. The anatomical and physiological attributes of the major brain noradrenergic nucleus, locus coeruleus (LC), that are reviewed in this chapter suggest that the LC–norepinephrine (LC–NE) system may integrate neuronal activity of anatomically and functionally diverse brain regions in response to external and internal stimuli. This chapter presents findings consistent with the hypothesis that activation of the LC–NE system by stressors serves a facilitatory role in the integration between endocrine, neuronal, and immunological limbs of the stress response and a pivotal role in the cognitive limb of the stress response. Data are also reviewed suggesting that some of the symptoms of certain stress-related psychiatric disorders, which the LC–NE system has been implicated in, may be extensions of the LC–NE response to stress.

II. Anatomical Characteristics of the LC–NE System

A. Widespread Efferent Projections

Although a detailed review of the anatomy of the LC–NE system is outside of the scope of this chapter (see Foote *et al.*, 1983 for review), certain anatomical features of this system highlight its potential as an integrator of the stress response. In the rat and primate, the LC is a relatively small, compact cluster of neurons that are homogenously noradrenergic (Swanson, 1976; Grzanna and Molliver, 1980), although other neurotransmitters may be colocalized with NE in LC neurons

0363–0951/98 $25.00

(Holmes and Crawley, 1995). A prominent feature of the LC is the widespread divergence of its projections (Ungerstedt, 1971; Swanson and Hartman, 1976; Foote *et al.*, 1983). Almost all of the norepinephrine in the forebrain originates from the LC, which densely innervates all areas of the cerebral cortex, the hippocampus, olfactory bulb, and thalamus (Anden *et al.*, 1966; Freedman *et al.*, 1975; Loy *et al.*, 1980; Foote *et al.*, 1983; Shipley *et al.*, 1985; Morrison and Foote, 1986). As for the forebrain, the LC is the sole source of NE in the cerebellum (Olson and Fuxe, 1971; Koda *et al.*, 1978; Foote *et al.*, 1983). Although the LC is not the sole source of NE in the hypothalamus, it projects to the medial part of the parvocellular paraventricular nucleus in the region of neurons that project to the median eminence and initiate the endocrine response to stress (Sawchenko and Swanson, 1982; Cunningham and Sawchenko, 1988). LC projections to brainstem nuclei and the spinal cord support a role for this system in sensory information processing (Fritschy *et al.*, 1987; Fritschy and Grzanna, 1990). Thus, the superficial layer of the dorsal horn of the spinal cord receives a dense innervation from the LC, as does the trigeminal nucleus (particularly the pars caudalis), which is involved in sensory information from the face (Fritschy *et al.*, 1987; Fritschy and Grzanna, 1990). In contrast, the LC does not innervate the intermediolateral column of the thoracic cord or autonomic brainstem nuclei, which receive norepinephrine innervation from other noradrenergic nuclei (Fritschy *et al.*, 1987; Fritschy and Grzanna, 1990), arguing against a direct role of the LC in autonomic function. Functions that have been attributed to the LC based on its efferent projections include arousal, memory and learning, and sensory processing (Foote *et al.*, 1983; Valentino and Aston-Jones, 1995). Electrophysiological studies have supported some of these functions (see discussion following).

Perhaps more important than its specific targets, is the remarkable divergence of this system to distant and functionally diverse regions of the central nervous system (CNS) (Foote *et al.*, 1983). Studies using multiple injections of retrograde tract tracers into different brain regions have revealed that individual LC neurons can project to the hippocampus and cortex, thalamus and cortex, thalamus and hippocampus, and to the forebrain and spinal cord (Ader *et al.*, 1980; Nagai *et al.*, 1981; Room *et al.*, 1981; Steindler, 1981; Foote *et al.*, 1983; Dietrichs, 1985). This structural divergence, and electrophysiological evidence for homogenous discharge patterns of LC neurons, support the idea that changes in LC discharge can simultaneously impact on functionally diverse CNS targets and that this could be a mechanism for coordinating the activity of multiple systems into a symptom complex.

It should be noted that in spite of the evidence for homogeniety of this system, it is becoming recognized that there is a topographical organization of LC efferents (Guyenet, 1980; Waterhouse *et al.*, 1983; Loughlin *et al.*, 1986a, b). The most obvious example of topographical specificity of efferents is the ventral location of spinally projecting LC neurons, which may also project to the cerebellum (Guyenet, 1980; Loughlin *et al.*, 1986a, b). The dorsal two-thirds of the LC project to the neocortex, hippocampus, hypothalamus, and cerebellum (Loughlin *et al.,* 1986a, b). A topographical organization of LC neurons projecting to different cortical regions has also been described (Waterhouse *et al.,* 1983). Topographically organized LC neurons with specific projections may also be distinguished morphologically (Loughlin *et al.,* 1986a, b). LC neurons sharing similar morphology and topography within the nucleus may also share other characteristics such as afferents, or colocalization of other neurotransmitters, and these characteristics may confer some functional specificity within the nucleus.

B. LC Afferents

Regarding its proposed role in the stress response, information about LC afferents is crucial to understanding the circuitry by which the LC becomes activated by different challenges. Definitive conclusions about all of the brain regions that directly impact on the LC is yet to be obtained, and this question is a topic of several recent anatomical studies. Part of the reason that this question remains unanswered to date involves technical limitations inherent in the anatomical tracers used to study LC afferents. Additionally, few of the afferents identified using anatomical techniques have been verified electrophysiologically. Finally, most studies using retrograde tract tracers to determine afferents to the LC have localized these tracers directly into the nucleus LC proper. It is now known that the LC has extensive dendritic aborizations that may extend for relatively long lengths into the dorsal pons (Shipley *et al.,* 1996). It is likely that afferent projections to these dendrites impact on LC physiology but are not identified in studies in which the tracer is injected into the nucleus LC proper (see below).

Early retrograde tract-tracing studies of LC afferents, which used horseradish-peroxidase (HRP), reported retrograde labeling in more than 30 nuclei, suggesting that afferent input to the LC was as diverse as its projection system (Aston-Jones *et al.,* 1991). In the past decade the develoment of tract tracers that remain more restricted after injection (wheat germ agglutinin [WGA]-conjugated HRP, WGA-apoHRP coupled to colloidal gold, the beta subunit of cholera toxin (CTb), and Fluoro-Gold) suggested that afferents to the nucleus LC are much more restricted than previously believed. Using these tracers and confirming results with electrophysiological findings, the nucleus paragigantocellularis (PGi) and nucleus prepositus hypoglossi (PrH) were identified as afferents to the LC (Aston-Jones *et al.,* 1986; Aston-Jones *et al.,* 1991). Other areas that were retrogradely labeled after injections of retrograde tracers into the LC included Barrington's nucleus, the dorsal cap of the paraventricular nucleus of the hypothalamus, the intermediate zone of the spinal cord, the nucleus Kölliker-Fuse, the periaqueductal gray, the lateral hypothalamus, and the preoptic area (Aston-Jones *et al.,* 1986; Guyenet and Young, 1987; Aston-Jones *et al.,* 1991; Ennis *et al.,* 1991; Rizvi *et al.,* 1994; Luppi *et al.,* 1995; Valentino *et al.,* 1996). Afferent projections to the LC from the PGi,

PrH, Barrington's nucleus, ventrolateral periaqueductal gray, the nucleus Kölliker-Fuse, and the medial preoptic area have been confirmed with anterograde tract tracing (Aston-Jones *et al.,* 1986; Aston-Jones *et al.,* 1991; Ennis *et al.,* 1991; Rizvi *et al.,* 1994; Luppi *et al.,* 1995; Valentino *et al.,* 1996).

As previously discussed, afferents to LC dendrites that extend outside of the nucleus can potentially impact on LC activity and may not be detected by retrograde tracing from the nucleus LC. A recent ultrastructural study demonstrated synaptic specializations between terminals that were anterogradely labeled from the central nucleus of the amygdala (CNA) and tyrosine hydroxylase immunoreactive dendrites in the rostrolateral pericoerulear region where LC dendrites extend (Van Bockstaele *et al.,* 1996a). Although CNA neurons are not retrogradely labeled from the LC, they are retrogradely labeled from this rostrolateral pericoerulear region, as are neurons of the paraventricular nucleus of the hypothalamus and bed nucleus of the stria terminalis (Van Bockstaele *et al.,* 1996c). These findings point toward limbic influences on LC activity through interactions in this pericoerulear region.

The neurochemical identity of some LC afferents has been determined in double-labeling studies using retrograde tract tracers and immunohistochemistry, and in some cases pharmacological studies have confirmed these findings. For example, putative glutamatergic neurons (identified by glutaminase-immunoreactivity) from the nucleus PGi project to the LC, and this glutamatergic input has been verified electrophysiologically (Forloni *et al.,* 1987; Toomin *et al.,* 1987; Ennis and Aston-Jones, 1988; Aston-Jones *et al.,* 1991; Drolet and Aston-Jones, 1991; Chiang and Aston-Jones, 1993). Substantial findings suggest that this input mediates the activation of LC neurons elicited by sciatic nerve stimulation (Ennis and Aston-Jones, 1988; Chiang and Aston-Jones, 1993) and opiate withdrawal (Rasmussen and Aghajanian, 1989; Rasmussen *et al.,* 1990; Akaoka and Aston-Jones, 1991). Other findings suggest that this input may also be important in LC activation by bladder distention (Page *et al.,* 1992, see below). It is noteworthy that this nucleus innervates both the LC and the preganglionic sympathetic nucleus in the spinal lateral horn, (Ross *et al.,* 1981; Milner *et al.,* 1988) and that sympathetic activity and LC discharge are often coactivated by the same stimuli (Svensson, 1987). The parallel activation of the sympathetic nervous system and the LC–noradrenergic system by a common afferent may be a mechanism for coordinating sympathetic activity and cognition. A similar parallel activation of the parasympathetic system and LC–noradrenergic system by Barrington's nucleus is described below.

The LC receives a rich enkephalinergic innervation, which derives from both the PrH and the PGi (Drolet *et al.,* 1992). Ultrastructural studies have demonstrated enkephalinergic synapses on LC neurons (Pickel *et al.,* 1979). Although this pathway has not been confirmed electrophysiologically, the potent effects of opiates on LC discharge activity (Williams *et al.,* 1982) are consistent with a physiologically relevant function of enkephalins in these LC projections.

Double-labeling studies using retrograde tracers and immunohistochemistry for phenethanolamine N-methyl transferase (a marker for adrenergic neurons) have demonstrated adrenergic projections from the PGi to the LC (Milner *et al.,* 1988; Pieribone and Aston-Jones, 1991). Electrophysiological studies suggest that this input may, in part, be responsible for the postactivation inhibition associated with LC responses to footshock (Aston-Jones *et al.,* 1991; Aston-Jones, Astier, Ennis, 1992).

Gamma amino butyric acid (GABAminergic) projections from the PrH to the LC have been confirmed in double-labeling studies and in electrophysiological studies involving electrical stimulation of the PrH (Mugnaini and Oertel, 1985; Ennis and Aston-Jones, 1989a, b; Aston-Jones *et al.,* 1991). The physiological significance of this pathway has yet to be determined.

With regard to stress, it is noteworthy that fibers immunoreactive for corticotropin-releasing hormone (CRH) have been demonstrated within the LC and in the pericoerulear region where LC dendrites extend (Swanson *et al.,* 1983; Valentino *et al.,* 1992). CRH is the hypothalamic neurohormone that initiates pituitary ACTH release in response to stress (Vale *et al.,* 1981). Recent ultrastructural studies revealed synaptic specializations between CRH-immunoreactive terminals and tyrosine-hydroxylase (TH) immunoreactive dendrites of LC neurons (Van Bockstaele *et al.,* 1996b) consistent with electrophysiological evidence for CRH-LC interactions (see below). In addition to direct synaptic contacts with LC dendrites, CRH-immunoreactive terminals contacted nonlabeled terminals that formed synapses with LC dendrites suggesting that they may affect LC activity by presynaptic modulation of other LC afferents (Figure 1).

Retrograde tract tracing from the nucleus LC combined with CRH-immunohistochemistry implicate the nucleus PGi, dorsal cap of the paraventricular nucleus of the hypothalamus, and Barrington's nucleus as sources of CRH input to the LC (Valentino *et al.,* 1992; 1996). Of these, Barrington's nucleus, which contains the greatest percentage of double-labeled neurons, is of great interest as a CRH afferent to the LC. Also known as the pontine micturition center, Barrington's nucleus projects to the sacral parasympathetic nucleus that innervates the bladder, distal colon, and genitals (Loewy *et al.,* 1979; Hida and Shimazu, 1982). Barrington's nucleus also projects to the dorsal motor nucleus of the vagus and so appears to be in a position to regulate autonomic parasympathetic activity (Valentino *et al.,* 1995). Retrograde tracing from both the LC and sacral spinal cord suggested that many Barrington's neurons, some of which are CRH-immunoreactive, diverge to project to both the LC and spinal cord (Valentino *et al.,* 1996). Thus, Barrington's nucleus could serve as a relay to coactivate the brain noradrenergic system and sacral parasympathetic system, thereby coordinating forebrain activity with pelvic visceral function. This may be analogous to the hypothesized parallel activation of the LC–noradrenergic system and autonomic sympathetic system by the nucleus PGi and may suggest anatomic substrates for linking the autonomic nervous system with the brain noradrenergic system (Figure 2). The implications of this for stress-related colonic disturbances is discussed later in the chapter.

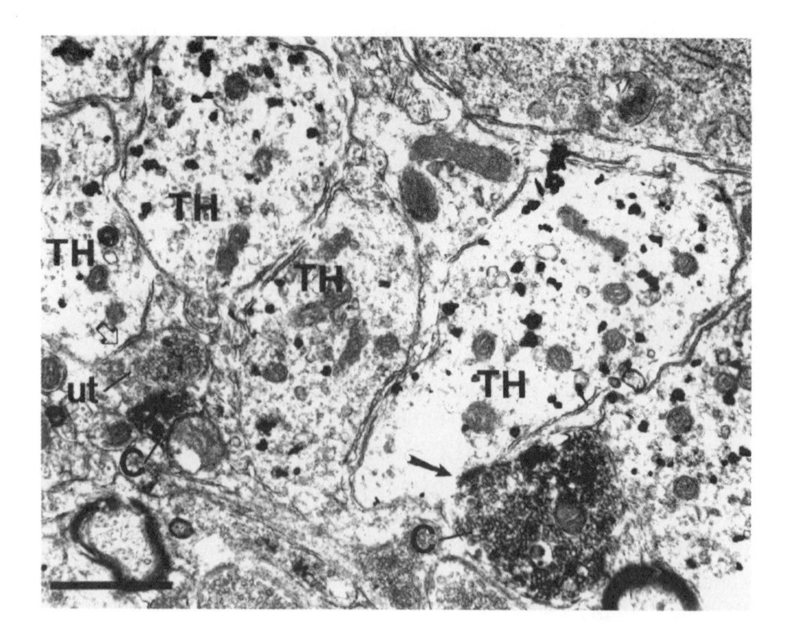

FIGURE 1 CRH-immunoreactive terminals (C) form asymmetric types of synaptic specializations with tyrosine-hydroxylase (TH)-immunoreactive dendrites in the LC. CRH and TH immunoreactivity were identified using immunoperoxidase and immunogold reactions, respectively. A CRH-labeled axon terminal forms an asymmetric contact (solid arrow) with a TH-labeled dendrite. Another CRH-labeled axon terminal is apposed to an unlabeled terminal (ut), which is in contact (open arrow) with a TH-labeled dendrite. Scale bar = 0.4 μm.

Anatomical studies also implicate the central nucleus of the amygdala (CNA) as a source of CRF that impacts on the LC. CRH neurons are numerous in the lateral CNA, and many of these are retrogradely from injections in the rostrolateral pericoerulear region into which LC dendrites extend (Van Bockstaele *et al.*, 1996c). Ultrastructural analysis revealed synaptic contacts between terminals anterogradely labeled from the CNA and LC dendrites (Van Bockstaele *et al.*, 1996a). Moreover, numerous terminals that were anterogradely labeled from the CNA were also CRH-immunoreactive (Van Bockstaele *et al.*, 1996c). These findings suggest that CRF from the CNA impacts on the LC via interactions with dendrites in the rostrolateral pericoerulear region. CRH neurons of the paraventricular nucleus of the hypothalamus and bed nucleus of the stria terminalis were also retrogradely labeled from this region, suggesting that CRH neurons in these nucleus may impact on the LC, although this has not been investigated at the ultrastructural level. Taken together with other data on CRF afferents to the LC to date, these results lead us to speculate that CRH inputs to the nucleus LC arise primarily from autonomic related nuclei, whereas CRH inputs to the rostrolateral dendritic region arise from limbic inputs that may convey information regarding emotional stressors to the LC (Figure 2).

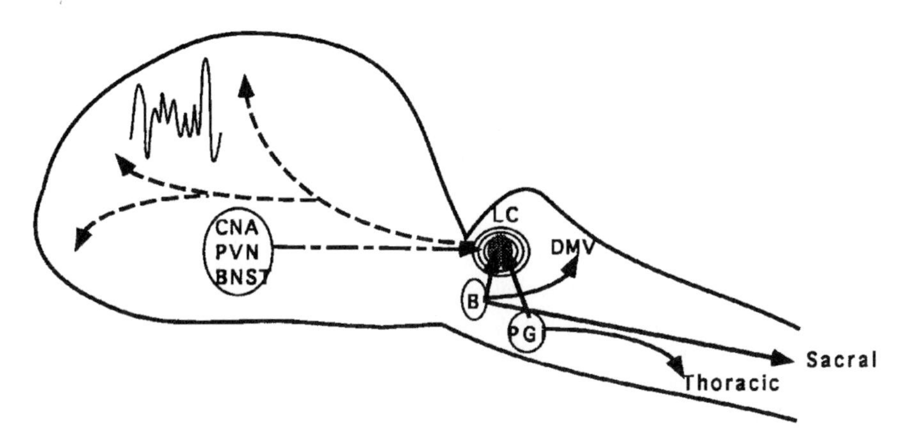

FIGURE 2 Schematic depicting putative CRH-afferents to the nucleus LC (gray oval) and pericoerulear region (indicated by concentric circles surrounding LC). Barrington's nucleus and the PGi project to the nucleus LC as well as to the autonomic parasympathetic (DMV and sacral spinal cord) or sympathetic nuclei (thoracic spinal cord), respectively. Parallel projections to the LC and autonomic nuclei may be an important means for integrating autonomic function with brain noradrenergic activity and its consequences (e.g., arousal, attention). In contrast to Barrington's nucleus and the PGi, limbic sources of CRH that could impact on the LC appear to terminate in the rostrolateral pericoerulear region (broken line). CRH from these sources may be important in relaying information regarding emotional stressors, as opposed to autonomic stressors. LC forebrain projections are indicated by dashed lines. Abbreviations: Barrington's nucleus (B); bed nucleus of the stria terminals (BNST); central nucleus of the amygdala (CNA); dorsal motor nucleus of the vagus (DMV); locus coeruleus (LC); paragigantocellularis (PGi); paraventricular nucleus of the hypothalamus (PVN).

III. Physiological Characteristics of LC–NE Neurons

A. State-dependent Neuronal Discharge

Much of the basis for the hypothesized functions of the LC–NE system have derived from electrophysiological studies of activity of these neurons in unanesthetized rats (Aston-Jones and Bloom, 1981a, b), cats (Jacobs, 1987), and monkeys (Foote *et al.,* 1980). A role for the LC in arousal was initially suggested by the finding that the discharge rate of LC neurons is well correlated to states of arousal as assessed by electroencephalographic (EEG) activity. In the cat, rat, and monkey LC neurons discharge infrequently during slow wave sleep and are silent during REM sleep (Foote *et al.,* 1980; Aston-Jones and Bloom, 1981a, b; Jacobs, 1987). Discharge rate is also relatively slow when animals are in a quiet waking state, for example, during grooming and eating. In contrast, discharge rate is relatively greater when animals are attending to external sensory stimuli (Foote *et al.,* 1980; Aston-Jones and Bloom, 1981a, b; Jacobs, 1987). Simultaneous recordings of the LC discharge, EEG, and electromyographic activity (EMG) suggest that changes in LC discharge precede cortical EEG activation and orienting motor behavior (Foote *et al.,* 1980; Aston-Jones and Bloom, 1981a, b; Jacobs, 1987).

Using selective, chemical manipulation of LC activity in anesthetized rats, Berridge and co-workers demonstrated that selective LC activation resulted in activation of the cortical and hippocampal EEG and that this was sensitive to systemic administration of propranolol (Berridge and Foote, 1991). Conversely, selective inactivation of the LC by local injections of clonidine resulted in high amplitude, slow wave EEG activity, similar to that observed in slow wave sleep (Berridge *et al.,* 1993). Together, these findings indicate that selective changes in the LC discharge rate are tanslated to an effect on forebrain electrophysiological activity. Moreover, they suggest that one function of LC activation may be to increase or maintain arousal. Indeed, this may be an important adaptive consequence of LC activation in stress (see below).

LC discharge rate covaries with the attentional state. In monkeys trained to respond to food reinforcement in an oddball visual discrimination paradigm, LC discharge rate was inversely correlated to pupillary fixation on a visual stimulus, which signaled a trial (Rajkowski *et al.,* 1992; 1994). When LC discharge rate exceeded a certain baseline level, pupillary fixation frequency decreased and performance in the task diminished. At this time, the monkeys appeared to be scanning stimuli in the environmental chamber that were not associated with the task. Rajkowski and coworkers hypothesized an inverted U-shaped relationship between performance on tasks requiring focused attention and LC discharge rate, whereby there is an optimal level of LC discharge which is associated with selective attention to stimuli integral to the task. At lower discharge rates, attention to all stimuli is diminished, and the animal appears drowsy. LC discharge rates greater than this optimal level would be predicted to be associated with decreased selective attention and attendance to multiple irrelevant stimuli in the environment. Thus, incremental changes in LC discharge rate are associated with shifts from inattention, to focused attention, to more scanning or labile attention. The relevance of this association between LC discharge and attention to the LC–NE role in the stress response is discussed later. It should be noted, however, that these studies are still correlational and await more direct evidence that changes in LC discharge rate signal attentional shifts.

B. Sensory Response of LC Neurons

One characteristic of LC neurons that has been a basis for hypothesizing a role for this nucleus in attention is their response to phasically presented sensory stimuli of different modalities. In unanesthetized animals (rats and monkeys) LC neurons are potently activated by short duration visual, auditory, olfactory, and nonnoxious somatosensory stimuli (Foote *et al.,* 1980; Aston-Jones and Bloom, 1981a). In the anesthetized state, only potentially noxious sensory stimuli (tail or paw pressure, sciatic nerve stimulation) effectively activate these neurons (Foote *et al.,* 1983). The sensory response of LC neurons is characterized by a brief (50 to 80 msec) activation followed by a longer duration of relatively attenuated activity. Glutamate efferents from the nucleus PGi mediate LC activation by sciatic nerve stimulation

(Ennis and Aston-Jones, 1988; Chiang and Aston-Jones, 1993), and preliminary findings implicate glutamate in the LC responses to auditory stimulation (Chiang *et al.,* 1991). Presumably, the brief burst in LC activity elicited by phasic sensory stimuli is translated to a burst of NE release in target regions, although this has not been demonstrated. Interestingly, under conditions of elevated tonic LC discharge, the sensory response of LC neurons appears attenuated. For example, stressors, such as CRH (Valentino and Foote, 1987; Valentino and Foote, 1988) and hypotension (Valentino and Wehby, 1988), and drugs, such as carbachol (Valentino and Aulisi, 1987), that increase LC discharge rate appear to decrease the signal-to-noise ratio of the LC sensory response. Similarly, in monkeys performing an oddball visual discrimination task, when tonic LC discharge is relatively high and focused attention is disrupted, the LC response to the target stimulus is attenuated (Rajkowski *et al.,* 1992; 1994). Thus, the LC sensory response may be associated with, or be an endpoint of, selective attention. When attention is more labile, the magnitude of this response is diminished.

In addition to sensory stimuli, LC neurons are activated by visceral and metabolic stimuli including hypotension, bladder or colon distention, hypoxia, and hypoglycemia (Svensson, 1987). Two neuromediators of LC activation by these stimuli include excitatory amino acids and CRH.

Like sciatic nerve stimulation and opiate withdrawal, LC activation by bladder distention appears to be mediated by glutamate afferents because it is completely prevented by intracerebroventricular (i.c.v.) administration of excitatory amino acid antagonists that are selective for non-N-methyl D aspartate (non-NMDA) type receptors and by local administration of kynurenic acid, a nonselective excitatory amino acid receptor antagonist, into the LC (Page *et al.,* 1992). The source of glutamate afferents mediating this response is unknown, but it may derive from the PGi because the pharmacological sensitivity of this response is similar to that for responses to sciatic nerve stimulation and opiate withdrawal, two stimuli which are thought to activate the LC via PGi glutaminergic inputs.

In contrast to bladder distention, colon distention and hypotension activate the LC via local release of CRH (Valentino *et al.,* 1991; Curtis *et al.,* 1994; Lechner *et al.,* 1997). Thus, CRH antagonists administered i.c.v. or intracoerulearly prevent LC activation by these stimuli, whereas glutamate antagonists are ineffective. Additionally, stimuli that selectively desensitize LC neurons to CRH (e.g., acute footshock stress) also desensitize LC neurons to activation by hypotension and colon distention (Curtis *et al.,* 1995; Lechner *et al.,* 1997). Conversly, CRH antagonists do not alter LC activation by bladder distention (Page *et al.,* 1992), opiate withdrawal (Valentino and Wehby, 1989), or sciatic nerve stimulation (Valentino *et al.,* 1991; Curtis *et al.,* 1994).

The functional consequences of LC activation by these various visceral and metabolic stimuli have not been determined. However, it is likely that one consequence is increased arousal. This has been demonstrated in studies where LC discharge and cortical and/or hippocampal EEG were simultaneously recorded during presentation of the physiological challenge (Valentino *et al.*, 1991; Page *et al.*, 1992; Page *et al.*, 1993). Colon and bladder distention and hypotension resulted in an activation of forebrain EEG that was temporally correlated with LC activation (Figure 3). Intracoerulear administration of clonidine (which inactivates the LC) or a CRH antagonist prevented EEG activation associated with hypotension. Similarly, kynurenic acid (administered i.c.v.), which prevented LC activation by bladder distention, also prevented the accompanying EEG activation (Page *et al.*, 1992). These results suggest that LC activation may be necessary in the arousal response to these stimuli.

C. Impact of the LC–NE System in Target Regions

The link between LC discharge and target cell effect, that is, NE release, has recently been examined using voltammetry or microdialysis studies. In general, these studies have demonstrated that stimuli that increase LC rate also increase NE levels in extracellular fluid, or increase the NE signal as measured by voltammetry, in postsynaptic targets (cortex, hippocampus, and thalamus) of the LC–NE system. These stimuli include electrical and chemical stimulation of the LC or dorsal noradrenergic bundle, footshock, restraint stress, hypotension, administration of CRH (both i.c.v. and intraLC), as well as administration of α_2 receptor antagonists

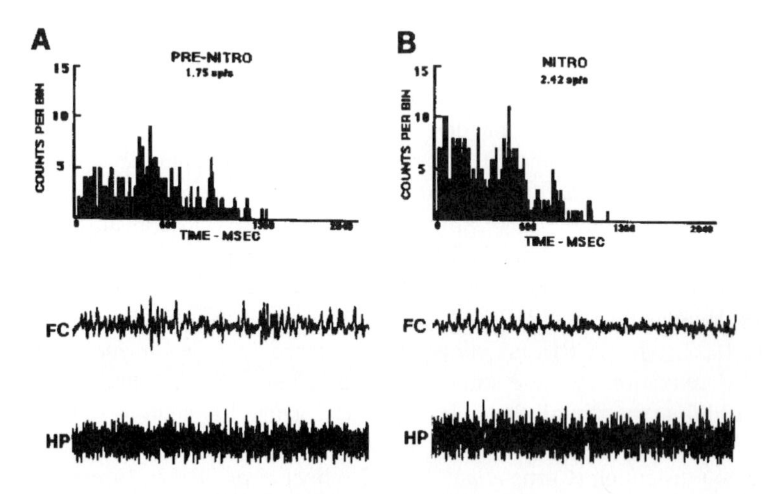

FIGURE 3 LC activation by hypotensive stress is associated with EEG activation recorded from the frontal cortex and hippocampus of the halothane-anesthetized rat. The top part of the figure shows interspike interval histograms of 3 min of LC discharge activity before (A) and 6 to 9 min after the initiation of a 15-min nitroprusside infusion (B). The mean discharge rate in this time period is indicated above the histograms. The bottom part of the figure shows the raw EEG traces recorded from frontal cortex (FC) and hippocampus (HP) that corresponded to the same times before (A) and during (B) nitroprusside infusion. See Valentino *et al.*, 1991, for details. Reprinted from Valentino *et al.* (1991) with kind permission of Elsevier Science-NL, Sara Burgerhartstraat 25, 1055 KV Amsterdam, The Netherlands.

(Heureux *et al.,* 1986; Abercrombie *et al.,* 1988; Brun *et al.,* 1991; Brun *et al.,* 1993; Lavicky and Dunn, 1993; Smagin *et al.,* 1994, 1995; Florin-Lechner *et al.,* 1996). Interestingly, NE-release-per-action potential may be greater when the same number of stimuli are presented as bursts, such as when LC discharge is evoked by phasic sensory stimuli, as opposed to evenly spaced, tonic stimuli (Figure 4).

The magnitude of NE release associated with LC discharge may be target specific. For example, in a study of the effects of visual stimulation on NE release (measured by voltammetry) in different cortical regions, it was demonstrated that NE release in monkeys' striate cortex exhibits an ocular dominance paralleling the ocular dominance for cortical neuron activation (Marrocco *et al.,* 1987). Moreover, in cats, visual stimuli that elicited NE release in the visual cortex failed to do so in the somatosensory cortex (Marrocco *et al.,* 1987). A possible explanation for the local specificity of NE release is that it may be regulated presynaptically by cortical afferents activated by specific stimuli. This type of presynaptic heteroregulation of NE release could be a mechanism for conferring specificity on the LC–NE system within terminal areas.

The impact of the LC–NE system on the electrophysiological activity of target neurons has been investigated in studies involving electrical stimulation of the LC or iontophoretic application onto target neurons. A detailed review of this literature is beyond the scope of this chapter, and readers are refered to Foote *et al.,* 1983. Early investigations into the postsynaptic effects of the LC–NE system suggested a generalized nonselective inhibition of target cells including cerebellar Purkinje neurons, hippocampal pyramidal cells, and neurons in the cortex, thalamus, and hypothalamus (Krynjevic and Phillis, 1963; Phillis *et al.,* 1967; Hoffer *et al.,* 1971; Hoffer *et al.,* 1973; Nakai and Takaori, 1974; Segal and Bloom, 1974a, b; Miyahara and Oomur, 1982; Foote *et al.,* 1983). Later studies revealed more complex effects of NE application that were concentration dependent; low doses selectively enhanced the effects of afferent inputs (evoked activity) relative to basal or spontaneous discharge, whereas higher doses resulted in general inhibition of all discharge (Foote *et al.,* 1975). These effects were observed in functionally diverse targets (i.e., cerebellum, cortex, and hypothalamus) and led to the idea that LC activation increases the signal-to-noise of activity of postsynaptic neurons (Foote *et al.,* 1983). Consistent with this, several studies demonstrated that iontophoresis of NE or LC stimulation potentiated the effects of both excitatory and inhibitory neurotransmitters on the same neuron (Freedman *et al.,* 1977; Moises *et al.,* 1979; Woodward *et al.,* 1979; Moises *et al.,* 1980; Waterhouse *et al.,* 1980; Moises *et al.,* 1981; Waterhouse *et al.,* 1982; Foote *et al.,* 1983). Taken together with the sensory-elicited activation of LC neurons, these findings suggested that the LC–NE system functions to facilitate processing of information about incoming sensory stimuli, as opposed to solely altering basal discharge rate.

The net effect of LC activation on a particular target neuron may depend on properties of the circuit in which the neuron functions. For example, one effect at-

A

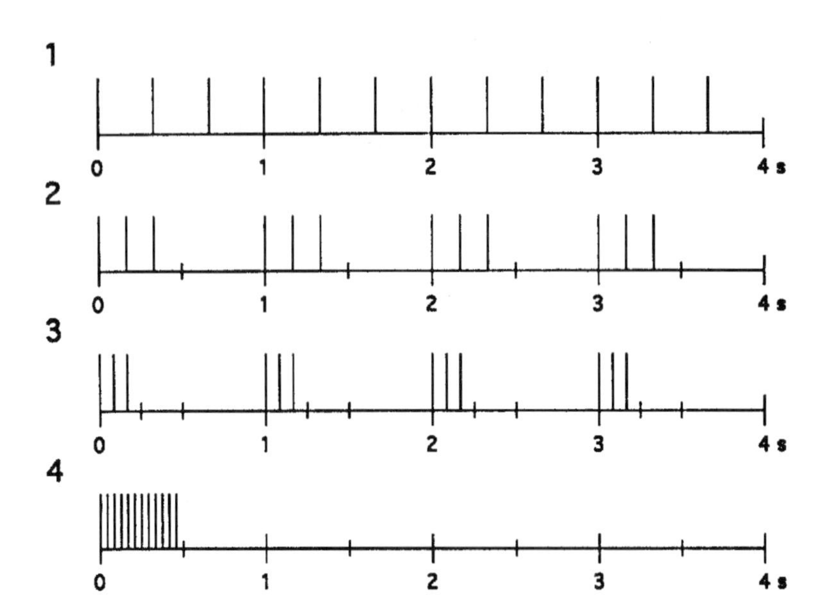
1
0
1
2
3
4 s
2
0
1
2
3
4 s
3
0
1
2
3
4 s
4
0
1
2
3
4 s

B

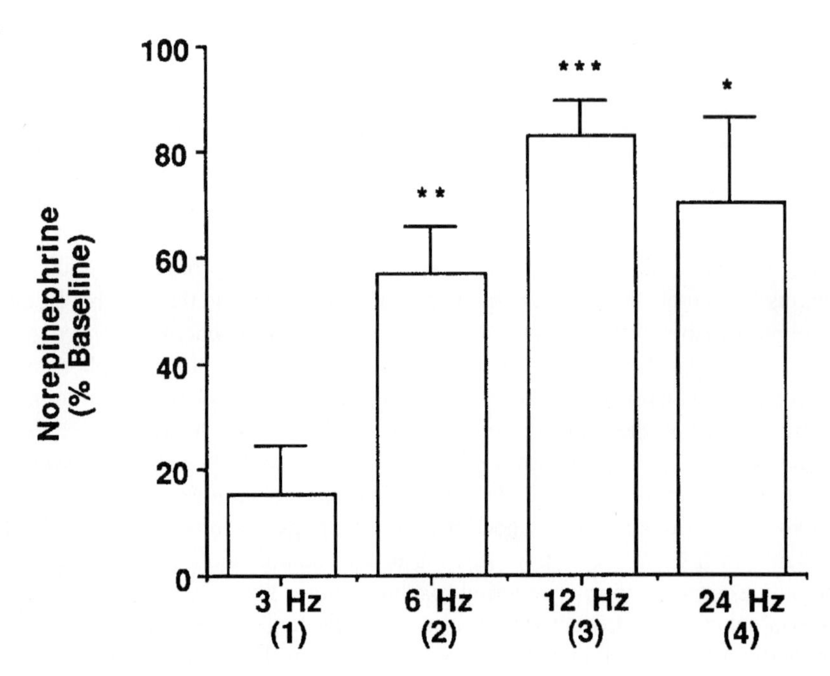
Norepinephrine
(% Baseline)
100
80
60
40
20
0
* *
* * *
*
3 Hz
(1)
6 Hz
(2)
12 Hz
(3)
24 Hz
(4)

tributed to NE in the visual cortex of the cat is to refine receptive fields (Moises *et al.,* 1990; Waterhouse *et al.,* 1990). NE application onto visual cortical neurons resulted in more sharply defined transitions between stimulus-induced inhibition and excitation. This effect would be predicted to enhance the ability to detect stimulus movement across receptive field boundaries. Another effect of NE and LC stimulation may be to alter patterns of neuronal firing. For example, in the thalamus NE can shift the pattern of neuronal activity from a "bursting" mode, which is associated with slow wave sleep and drowsiness, to a single spike firing mode, which is associated with the transmission of sensory stimuli to the cerebral cortex and waking and attention (McCormick and Prince, 1988). This pattern shift may underlie, in part, LC effects on arousal.

Importantly, these studies suggest that the impact of the LC–NE system on target neurons may be more complex than previously thought and may not readily generalize from one target region to another.

IV. Stress Effects on Brain Noradrenergic Function

A. NE Release and Turnover in Target Regions

The potent activation of the LC–NE system by a variety of stressors has been a strong basis for initially hypothesizing a role for this system in stress. This was demonstrated in studies measuring the neurochemical output of the LC–NE system in target regions and in electrophysiological studies of the LC neuronal discharge.

Figure 4 Comparison of the effects of phasic vs. tonic electrical LC stimulation on norepinephrine level in the medial prefrontal cortex. (A) Schematic representation of LC stimulation parameters. Stimulation patterns were designed to deliver the same number of pulses in a 20-min period (3,600). Each vertical line represents a stimulus (700 μA, 0.2 ms). Panel 1 represents tonic 3 Hz stimulation, with 3 stimuli delivered in an evenly spaced pattern each second. Panel 2 represents stimulation with 6 Hz trains (500 ms duration), with 3 stimuli delivered in the first 500 ms of each second. Panel 3 represents stimulation with 12 Hz trains (250 ms duration), with 3 stimuli delivered in the first 250 ms of each second. Panel 4 represents stimulation with 24 Hz trains (500 ms duration), with 12 stimuli delivered in the first 500 ms of each 4 seconds. (B) Bars represent the mean norepinephrine level measured in the 20 min stimulation sample expressed as a percentage of the mean baseline level determined from the 3 samples prior to stimulation. Vertical lines represent 61 Standard error of the mean (S.E.M.). Norepinephrine levels evoked by all three phasic stimulation patterns of 6 Hz (n 5 6), 12 Hz (n 5 6), and 24 Hz (n 5 5) were significantly greater than levels evoked by 3 Hz tonic stimulation (n 5 6). Asterisks indicate *$p < 0.05$, **$p < 0.01$, ***$p < 0.001$; Scheffe test for post hoc comparisons. In addition, phasic stimulatio with 12 Hz further increased norepinephrine levels above levels evoked by phasic stimulation with 6 Hz ($p < 0.05$). Levels of norepinephrine evoked by phasic 24 Hz stimulation were no different than levels evoked by phasic stimulation with 6 or 12 Hz. (A) Reprinted from "Enhanced norepinephrine release in prefontal cortex with burst stimulation of the locus coeruleus" by S. M. Florin-Lechner, J. Druhan, G. Aston-Jones, and R. J. Valentino in *Brain Research* **742:** 89–97, copyright © 1996 by Elsevier, with kind permission of Elsevier Science-NL, Sara Burgerhartstraat 25, 1055 KV Amsterdam, The Netherlands.

The earliest studies demonstrating activation of the LC–NE system by stress involved measurements of NE metabolites and turnover in different brain regions during stress (Thierry *et al.*, 1968; Korf *et al.*, 1973; Cassens *et al.*, 1980). Selective increases in NE turnover were observed in the brainstem, mesencephalon, and spinal cord in rats exposed to footshock. In addition, footshock increases NE turnover in the cerebral cortex and hippocampus, two regions that receive their sole NE input from the LC. Unilateral LC lesions abolished these effects on the ipsilateral side. Finally, these increases in NE turnover were also elicited by stimuli conditioned to footshock.

Recently, more direct methods for measuring NE release in LC target regions, such as microdialysis, have yielded results consistent with turnover studies. Thus, restraint, tailshock, and tailpinch increase extracellular NE levels in the hippocampus in unanesthetized rats (Abercrombie *et al.*, 1988). Auditory stress also increased NE release in the cortex (Britton *et al.*, 1992), and the same hypotensive stress (elicited by intravenous [i.v.] nitroprusside infusion) that was previously demonstrated to increase LC discharge rates, also increased NE release (measured by microdialysis) in the frontal cortex and hypothalamus (Smagin *et al.*, 1994).

B. Stress Effects on LC Neurons

Because some of the brain regions previously discussed receive their sole source of NE from the LC, it is likely that activation of LC cell bodies by these stressors is responsible for the change in neurotransmitter output. Consistent with this, most of the challenges that increase NE turnover or release have also been demonstrated to increase LC discharge rate in rats and cats. These include restraint, shock, tailpinch, auditory stress, hypotension, and hemmorage (Abercrombie and Jacobs, 1987; Morilak *et al.*, 1987a, b; Svensson, 1987; see also for review Valentino *et al.*, 1993). In support of electrophysiological findings of increased LC discharge rates by stressors, other indices of activation of LC cell bodies by stress have been reported, including increased expression of tyrosine hydroxylase (Melia and Duman 1991; Melia *et al.*, 1992) and increased expression of c-fos (determined by immunoreactivity) in LC neurons of rats exposed to footshock (Valentino *et al.*, 1992).

Taken together, the neurochemical and electrophysiological studies described provide firm evidence that the LC–NE system is activated by stress. However, the circuitry and neurotransmitters underlying this activation were unknown. Two potential candidates for LC activation by stressors include glutamate and CRH. Glutamate has been demonstrated to mediate LC activation by a number of stimuli including sciatic nerve stimulation, opiate withdrawal, and bladder distention (see previous discussion). However, although these are arousing stimuli, it is not clear whether they can be considered stressors in the classical sense, that is, whether they elicit the appropriate endocrine response. Immunohistochemical findings (described previously) suggesting that the stress-neurohormone, CRH, may communicate with the LC, led to studies designed to test the hypothesis that CRH is a mediator of LC activation by stressors.

C. CRH-LC Link in Stress

It is noteworthy that prior to the investigations to be described, substantial anatomical, physiological, and behavioral studies supported the idea that CRH could serve as a neurotransmitter in extrahypophyseal brain regions (Valentino *et al.,* 1993). For example, numerous CRH-containing cell groups were observed in the brain outside of the hypothalamic-pituitary axis, and CRH-immunoreactive (CRH-IR) fibers were visualized in nuclei that were not directly involved with endocrine aspects of stress responses (Cummings *et al.,* 1983; Swanson *et al.,* 1983; Sakanaka *et al.,* 1987; Valentino *et al.,* 1992). Intracerebroventricular administration of CRH was demonstrated to alter cardiovascular function (Fisher *et al.,* 1982; Brown and Fisher, 1985), gastrointestinal function (Tache *et al.,* 1983; Tache and Gunion, 1985), and behavior (Sutton *et al.,* 1982; Britton *et al.,* 1985; Kalin, 1985; Eaves *et al.,* 1985), even in hypophysectomized or adrenalectomized animals, indicating that these effects are independent of the neurohormone action of CRH. Taken together, these studies suggested that CRH may have a global role in the stress response, serving as a neurotransmitter in regions outside of the hypothalamic-pituitary axis to mediate behavioral and/or autonomic responses to stress in addition to participating as a neurohormone in hypophyseal circuits to initiate the cascade of endocrine effects that has been considered to be the hallmark of stress. Studies demonstrating that CRH antagonists prevent many stress-elicited responses, including cardiovascular effects, gastrointestinal changes, and several behavioral responses, are consistent with this hypothesis (Brown *et al.,* 1985; Britton *et al.,* 1986; Tazi *et al.,* 1987; Kalin *et al.,* 1988; Lenz *et al.,* 1988; Stephens *et al.,* 1988).

Initial studies examining CRH-LC interactions demonstrated that CRH, administered i.c.v. in doses that mimic other aspects of the stress response (3 to 10 μg), increased spontaneous LC discharge rates of anesthetized rats (Valentino *et al.,* 1983), and CRH was somewhat more potent and efficacious when administered i.c.v. to unanesthetized rats (Valentino and Foote, 1988). LC activation by i.c.v.-administered CRH was likely due to direct actions within the nucleus because intracoeruleuar injection of CRH antagonists prevented LC activation by i.c.v.-administered CRH (Curtis *et al.,* 1994, in press; Valentino *et al.,* 1983), and the antagonists were 300 to 1,000 more potent when administered directly into the LC (Curtis *et al.,* 1994). Consistent with a local effect of CRH within the LC, intracoerulear CRH increased LC discharge rate and CRH was 200 times more potent when administered directly into the LC vs. i.c.v. (Curtis *et al.,* 1997). Intracoerulear CRH also increased NE release in the cortex and hippocampus as measured by microdialysis (Smagin *et al.,* 1995; Page *et al.,* 1995; Curtis *et al.,* 1997). Although these studies indicate an excitatory effect of CRH on LC neurons with a site of action within the LC, these effects have yet to be demonstrated in vitro. The ultrastructural demonstration of CRH interactions with LC dendrites and terminals of LC afferents support either direct or indirect modulatory actions of CRH on LC discharge (Van Bockstaele *et al.,* 1996b). Studies reporting a lack of mRNA for a CRH receptor in LC neurons argue against direct interactions between CRH and LC neurons (Wong *et al.,* 1994; Potter *et al.,* 1994). However, the results of these *in situ* studies should also be interpreted with caution because probes used to measure the CRH receptor message may lack the appropriate sensitivity for measurement of levels in LC neurons and because CRH receptors may be transported to dendrites that extend away from the nucleus.

CRH also alters LC responses to sensory stimuli by elevating baseline discharge and attenuating evoked discharge thereby reducing the signal-to-noise ratio of the LC sensory response (Valentino and Foote, 1987; 1988; Valentino and Wehby, 1988). This effect has been observed in anesthetized rats using sciatic nerve stimulation to evoke LC discharge (Valentino and Foote, 1987; Valentino and Wehby, 1988) and in unanesthetized rats where auditory stimuli were used (Valentino and Foote, 1988). Whether these effects are due to presynaptic actions of CRH on terminals of LC afferents or postsynaptic effects on LC neurons has not been determined. The effects of CRH on LC sensory responses may play a role in shifts in the attentional state (see the following discussion).

The finding that exogenously administered CRH activates LC neurons, taken together with ultrastructural evidence for interactions between CRH terminals and LC dendrites, suggests that endogenously released CRH may mediate activation of LC neurons by stressors. This hypothesis was first tested with hypotensive challenge produced by an intravenous infusion of nitroprusside (Valentino and Wehby, 1988), which causes a 50% decrease in mean arterial blood pressure. Hypotensive stress mimicked the neuronal effects of CRH on LC discharge, that is, it increased the spontaneous discharge rate and decreased the signal-to-noise ratio of the LC sensory response (Valentino and Wehby, 1988). Similarly, removal of 2 ml of blood from the arterial system also increased LC discharge by a similar magnitude (Valentino and Wehby, 1988). These findings are in agreement with earlier reports (Svensson, 1987). Interestingly, in unanesthetized rats, higher concentrations of nitroprusside are required to produce comparable magnitudes of hypotension, perhaps because anesthesia inhibits systems that counteract hypotension (Curtis *et al.,* 1993). However, in unanesthetized rats, the magnitude of LC activation elicited by a comparable level of hypotension is greater compared to anesthetized rats (Curtis *et al.,* 1993). This is consistent with the finding that CRH is more potent in activating the LC of unanesthetized rats compared to anesthetized rats (Valentino and Foote, 1988).

CRH antagonists (administered i.c.v.), but not glutamate antagonists, blocked LC activation by hypotension in a dose-dependent manner, and the potencies for an individual antagonist (either α helical $CRH_{9\text{-}41}$ or D-Phe $CRH_{12\text{-}41}$) were identical against CRH and hypotensive challenge, indicating that similar receptor binding sites are involved in LC activation by these two stimuli (Valentino *et al.*, 1991; Curtis *et al.*, 1994). I.c.v. administration of α helical $CRH_{9\text{-}41}$ was also demonstrated to attenuate LC ativation by hypotensive challenge in unanesthetized rats (Curtis *et al.*, 1993). Importantly, intracoerulear microinjection of either D-Phe $CRH_{12\text{-}41}$ or α helical $CRH_{9\text{-}41}$ prevented LC activation by hypotensive stress, and the antagonists were approximately 300–1,000 times more potent when administered directly into the LC compared to i.c.v. (Curtis *et al.*, 1994). These convergent findings strongly support the hypothesis that CRH release into the LC mediates its activation by hypotensive stress. Finally, systemic administration of dexamethasone, in doses that prevent CRH release by hypotensive challenge, did not attenuate LC activation suggesting that this effect of CRH is independent of its neurohormone role (Valentino and Wehby, 1988). Consistent with these electrophysiological studies, microdialysis studies demonstrated that this same hypotensive stress elicits NE release in the prefrontal cortex, which receives all of its NE from the LC, and in the hypothalamus (Smagin *et al.*, 1994).

Analogous studies demonstrated that LC activation by low magnitudes of colon distention require CRH release in the LC (Lechner *et al.*, 1997). Thus, this activation is greatly attenuated by intracoerulear administration of CRH antagonists. Additionally, stimuli that selectively desensitize LC neurons to CRH dramatically decrease the magnitude of LC activation by colon distention (Lechner *et al.*, 1997).

The circuits underlying LC activation by these stimuli have yet to be elucidated. As discussed previously, potential CRH afferents to the nucleus LC include Barrington's nucleus, the nucleus PGi, and the dorsal cap of the paraventricular nucleus of the hypothalamus. CRH afferents to the rostrolateral pericoerulear region overlapping LC processes include the amygdala, paraventricular nucleus of the hypothalamus, and the bed nucleus of the stria terminalis. With regard to hypotensive challenge, it is tempting to speculate that CRH neurons in the PGi mediate this response because this nucleus receives projections from the nucleus of the solitary tract which in turn receive baroreceptor input, and the PGi projects to the intermediolateral column of the spinal cord (Ross *et al.*, 1981). As hypotensive challenge has been demonstrated to activate PGi neurons (Brown and Guyenet, 1986), it is tempting to speculate that it activates CRH neurons within the PGi that project to the LC.

With regard to colon distention, Barrington's nucleus is a likely source of CRH that would impact on the LC. CRH neurons of Barrington's nucleus have been identified that project to both the LC and sacral spinal cord (Valentino *et al.*, 1996), and because Barrington's nucleus has been associated with pelvic visceral function, it is likely that these neurons are affected by colonic stimuli. This could result in coactivation of both the LC and sacral parasympathetic system to produce a coordinated cognitive and autonomic response to the distention. The role of this hypothesized system in stress-related colonic disorders is discussed below.

V. Adaptive Consequences of LC–NE Activation by Stress

Although substantial evidence indicates that the LC–NE system is activated by stressors, the consequences of activating this system have remained unknown. This is of particular interest with challenges such as hypotension, which produce a relatively small magnitude of neuronal activation, that is, 30 to 40% increase in anesthetized animals and 100% in unanesthetized animals. Although this effect appears to be translated to increased NE release in LC target regions (Smagin *et al.*, 1994), it is important to verify some functional consequences of this effect. The following discussion addresses some potential adaptive consequences of activation of the LC–NE system by stressors.

A. Forebrain Arousal

As previously discussed, the electrophysiological characteristics of LC–NE neurons *in vivo* have implied a role for this system in arousal. Particularly, the studies of Berridge and colleagues (Berridge and Foote, 1991; Berridge *et al.*, 1993) indicate that selective manipulation of LC discharge can dramatially affect the cortical and hippocampal EEG. Based on these studies, they hypothesized that one consequence of LC activation by stressors was increased arousal. To test this hypothesis, cortical and/or hippocampal EEG activity were recorded simultaneously with LC neuronal activity during physiological challenges. When LC discharge was increased by hypotension or bladder distention, both the cortical and hippocampal EEG became activated (Valentino *et al.*, 1991; Page *et al.*, 1992; Page *et al.*, 1993) (See Figure 3). This was apparent as a decrease in large amplitude, slow wave activity in the cortical EEG and a shift from mixed frequency activity to theta rhythm recorded from the hippocampus. LC activation by colon distention was also temporally correlated with activation of cortical EEG (Lechner *et al.*, 1997). Glutamate antagonists (administered by i.c.v.), which prevented LC activation by bladder distention, also prevented the EEG activation associated with bladder distention (Page *et al.*, 1992). Selective inactivation of the LC produced by microinjection of clonidine into the LC prevented both LC activation by hypotensive stress and the accompanying EEG activation (Page *et al.*, 1993). Finally, local administration of CRH antagonists into the LC (which do not inhibit the LC spontaneous discharge rate) prevented LC activation by hypotensive challenge and the associated EEG activation (Page *et al.*, 1993). These results indicate that even relatively small increases in LC discharge rate (on the order of 30 to 40%) can be translated to forebrain EEG activation. Moreover, the results suggest that one consequence of CRH activation of the LC during stress may be to increase or maintain arousal. Indeed, this may be adaptive in a naturalistic situation (e.g., if an animal has been attacked by a predator and is hemorrhaging, it is adaptive to maintain arousal to escape). As discussed later, this same consequence could be pathological if it persisted or occurred inappropriately. Based on the finding that LC activation by bladder distention is also associated with EEG activation (Page *et al.,* 1992) and the studies of Berridge and colleagues (Berridge and Foote, 1991; Berridge *et al.,* 1993), it is likely that increased arousal is a common consequence of LC activation by numerous stimuli (including those that may not be classified as stressors).

B. Disruption of Focused Attention

Studies by Rajkowski (1992, 1994) and colleagues in monkeys performing an oddball visual discrimination tasks demonstrated that relatively small increases in LC discharge rates (30%) were associated with a disruption of performance on tasks requiring focused attention (Rajkowski *et al.,* 1992; 1994). During the period that LC discharge was increased and performance on the task was poor, the animals appeared to be attending to multiple stimuli in the behavioral chamber that were not relevant to the task. Moreover, the response of LC neurons to the target stimulus was greatly attenuated during this peiod. This group hypothesized that increases in LC discharge rate (above some optimal level) are associated with a shift from focused to scanning or labile attention. The increases in LC discharge rate that are associated with these shifts in attention are comparable to the increases produced by hypotensive stress and i.c.v. administration of CRH. Although a causal relationship between LC discharge and shifts in attention have yet to be demonstrated, these findings suggest that LC activation produced by stressors may be associated with a shift from focused to scanning or labile attention. The observation that both hypotensive stress and CRH attenuate LC sensory responses (Valentino and Foote, 1987; Valentino and Foote, 1988; Valentino and Wehby, 1988), similar to the attenuation of LC responses to the target stimulus observed during states of scanning attention, is consistent with this hypothesis. A change from focused to scanning attention may be adaptive in certain naturalistic situations. However, as previously discussed for increased arousal, this consequence could be pathologic if it persisted or if it were elicited inappropriately.

C. Behavioral Responses to Stress

The LC–NE system has been implicated in several behavioral responses to stressors based on different paradigms. For example, CRH enhances conditioned fear; this effect is specifically antagonized by systemic administration of 1-propranolol, indicating that the interaction of NE with B-receptors is integral to this effect (Cole and Koob, 1988). Similar results were obtained in experiments measuring CRH-induced defensive withdrawal (Dunn and Berridge, 1987). A more specific role for CRH in the LC in shock-induced freezing has been suggested by studies demonstrating that CRH antagonists administered into the LC reduce the duration of freezing (Swiergel *et al.,* 1992). Microinjection of CRH into the LC also increased anxiogenic behaviors in an open field paridigm, and this injection was associated with significant increases in the concentration of the norepinephrine metabolite 3,4-dihydroxyphenylglycol in forebrain regions (Butler *et al.*, 1990). Taken together, these studies suggest that LC activation by CRH during stress may be important for some of the adaptive behavioral responses to stress. A recent report examining the effects of CRH on retention of a passive-avoidance task suggests that CRH activation of the LC may also be important in memory retention (Chen *et al.*, 1992). In this study microinjections of CRH directly into the LC enhanced retention latency, and this effect was prevented in rats treated with 6-hydroxydopamine (6-OHDA). Doses of CRH that improved retention were associated with an increase in NE in the hippocampus. Interestingly, the effects of CRH were mimicked by yohimbine, which also increases the LC discharge rate.

D. Gastrointestinal Responses to Stress

Recent studes implicate the LC in stress-induced increases in colonic motility. Both stress (restraint, conditioned fear) and central administration of CRH increase colonic transit and fecal output (Tache *et al.*, 1993). Microinjections of CRH into the paraventricular nucleus of the hypothalamus and LC mimicked the effects of i.c.v.-administered CRH (Monnikes *et al.*, 1992a, b). In contrast, microinjections of CRH into the lateral hypothalamus or amygdala were ineffective (Monnikes *et al.*, 1992b). Although the LC (particularly its ventrally located neurons) sends descending projections to the spinal cord, it is not clear what pathways would mediate increased colonic motility by CRH microinjection into the LC. As previously discussed, based on the pattern of its spinal terminations, the LC–NE system is more likely to be involved in processing of sensory information than in autonomic function. However, communications between the LC and the nearby Barrington's nucleus, which projects to the sacral parasympathetic nucleus, could be a possible mechanism whereby intracoerulear CRH could elicit changes in colonic motility.

E. Immune Response to Stress

An important component of the stress response involves the inhibition of immune function, which occurs at several levels. Glucocorticoids are well known as inhibitors of natural killer cells and lymphocyte activity. However, the central CRH also inhibits the function of natural killer cells and lymphocytes through activation of the sympathetic nervous system, independent of its activation of the pituitary-adrenal axis (Strausbaugh and Irwin 1992; Irwin, 1993). The immunosuppressant effects of the hypothalamic-pituitary-adrenal (HPA) response to stress are thought to be part of a feedback loop because cytokines, such as interleukin-1, which are released from macrophages on immune challenge, elicit hypothalamic CRH release and activation the HPA axis (Sapolsky *et al.*, 1987; Uehara *et al.*, 197; Berkenbosch *et al.*, 1987). The noradrenergic projection to the hypothalamus has been implicated in interleukin-1 activation of the HPA axis because increases in plasma corticosterone produced by interleukin-1 are accompanied by increased norepinephrine turnover in the hypothalamus (Dunn, 1992). Moreover, the 6-OHDA lesion of the ventral noradrenergic bundle, which is the major NE projection to the hypothalamus, prevents HPA activation (measured by plasma corticosterone) elicited by interleukin-1 (Chuluyan *et al.*, 1992).

A recent study demonstrating that local administration of CRH into the LC decreases T-lymphocyte mitogenic responses suggests that the LC may be an important link in stress-immune interactions (Rassnick *et al.*, 1994). The mechanism by which LC activation leads to immunosuppression is not clear. Stress- and CRH-induced suppression of lympocyte function requires sympathetic nervous system activation because it is prevented by transection of the splenic nerve and by B-adrenergic receptor antagonists (Cunnick *et al.*, 1990; Irwin *et al.*, 1990; Wan *et al.*, 1993). However, LC activation does not consistently increase sympathetic activity (Sved and Felsten, 1987; Miyawaki *et al.*, 1991). Certain findings suggest that LC effects on immune function are a result of HPA activation. Thus, decreases in the T-lymphocyte function produced by CRH injections into the LC were associated with increased plasma ACTH and corticosterone (Rassnick *et al.*, 1994). Taken together with the findings previously described, a feed-forward circuit may be hypothesized whereby stress elicits CRH release in the LC which then contributes to HPA activation and a suppression of the immune response. Nonetheless, the finding that CRH-induced suppression of the immune function can be dissociated from activation of the pituitary-adrenal axis (Irwin, 1993) suggests the involvement of other extrahypophyseal regions in this response. The LC may be an important part of this circuit. These findings, implicating a role for the LC in the immune limb of the stress response, are intriguing; they suggest that the LC may link neural and immune responses as well as endocrine and cognitive limbs of the stress response.

VI. Pathological Consequences of Noradrenergic Activity in Stress

An important characteristic of the endocrine stress response is its feedback regulation, which limits an adaptive response that would otherwise be harmful if it persisted. In spite of this feedback regulation, the HPA response to stress is facilitated in animals with a history of repeated or chronic stress (Dallman *et al.*, 1992). Evidence suggests that the mechanism for facilitation occurs at, or above, the level of the hypothalamus and results in increased drive to hypothalamic CRH neurons, which is insensitive to glucocorticoid feedback (Dallman *et al.*, 1992). Increased activity of the neurohormone CRH would be expressed as hypercortisolemia and altered sensitivity to glucocorticoids. Indeed, this is characteristic of stress-related psychiatric disorders, such as depression. Facilitation of the neurotransmitter actions of CRH could play a role in nonendocrine symptoms of stress-related psychiatric disorders. This has been examined in investigations of the regulation of CRH-LC interactions.

A. LC Activity in Animals with a History of Stress

As previously discussed, activation of the LC–NE system by stressors has direct behavioral, cognitive, gastrointestinal, and immunity consequences that may be adapative in the acute response to a stressor. However, as is the case for other aspects of the stress response, these consequences could be pathological if they persisted or were elicited inappropriately. Thus, hyperarousal, difficulty in focusing attention, anxiogenic behaviors, and changes in gastrointestinal motility are adaptive short-term responses that are also characteristic symptoms of stress-related psychiatric disorders. It is possible that repeated or chronic stress, which sensitizes animals to the neurohormone activity of CRH, produces a parallel facilitation of the CRH neurotransmitter function in the LC. This hypothesis was recently tested in rats exposed to acute or repeated inescapable footshock (Curtis *et al.,* 1995). Neither acute or repeated prior stress altered the LC spontaneous discharge rate or LC responses to repeated sciatic nerve stimulation (mediated by glutamate inputs to LC). However, prior stress had a dramatic influence on LC sensitivity to CRH (Figure 5). In rats with a history of acute prior stress, LC neurons were no longer activated by hypotensive challenge, and the response of LC neurons to 3-μg dose of CRH (i.c.v.) was greatly attenuated. These results suggested that acute stress results in a rapid postsynaptic desensitization of LC neurons to CRH. In repeatedly stressed rats, LC responses to the same dose of CRH were similarly diminished; however, the response to hypotensive challenge was reestablished. Initially, these results suggested that the same challenge released more CRH in repeatedly stressed rats, perhaps in order to compensate for decreased postsynaptic sensitivity. However, further examination of CRH dose-response curves in these animals provided an alternative explanation (see Curtis *et al.,* 1995). In rats exposed to one session of shock, the CRH dose-response curve for LC activation was shifted to the right, again indicative of postsynaptic desensitization. In repeatedly stressed rats, although the maximum magnitude of LC activation produced by CRH was decreased (by approximately 50%), the dose-response curve was shifted to the left such that low doses of CRH, which are usually below threshold for activating LC neurons, produced a significant elevation of the LC discharge rate. Importantly, the magnitude of LC activation produced by these previously subthreshold doses was comparable to that necessary to increase forebrain EEG activity (Page *et al.,* 1993), that is, sufficient to impact on forebrain targets. The shifts in the CRH dose-effect curve observed in acutely and repeatedly stressed rats could account for the loss, and the return, of LC responses to hypotensive stress in the different groups, respectively.

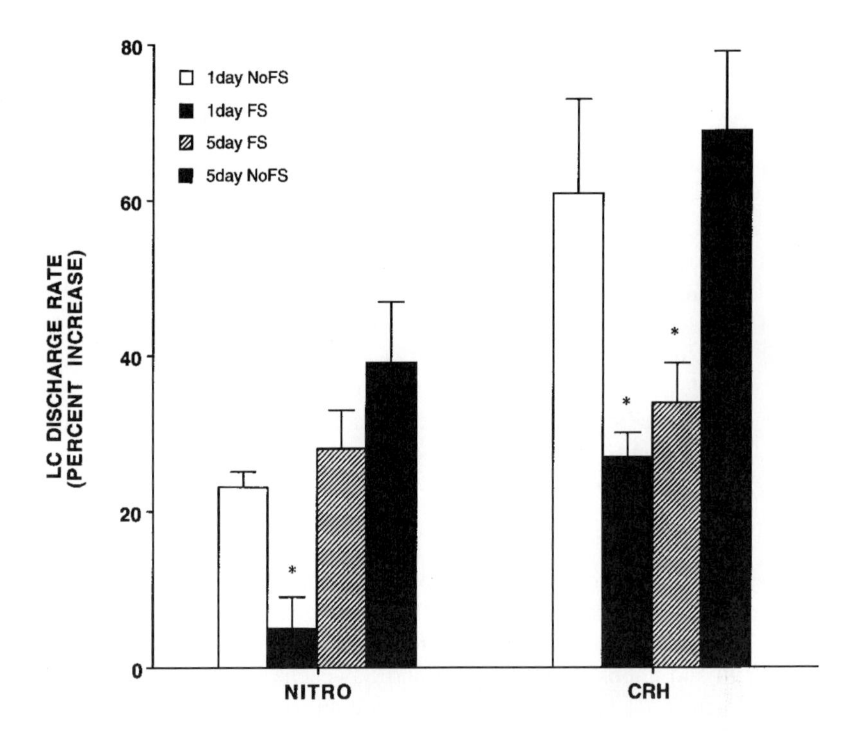

FIGURE 5 Effect of prior stress on LC activation by endogenous and exogenous CRH. Bars represent the mean percentage increase in LC discharge produced by CRH (3.0 μg, i.c.v.) or hypotensive challenge in acutely stressed rats (1 day No FS [footshock]; open bars), acute matched controls (1 day FS; gray bars), repeatedly stressed rats (5 day FS; hatched bars), or controls matched for repeated stress (5 day NoFS; solid bars). Vertical lines represent 61 S.E.M. Acute and repeated stress attentuate the response to exogenous CRH (3.0 μg) by about 50%. However, the response to hypotension (which is mediated by endogenous CRH) is attneuated only in acutely stressed rats. For details see Curtis *et al.* (1995).

The results of the study suggest that sensitization of LC neurons to CRH in repeatedly stressed rats may be a mechanism allowing the LC–NE system of chronically stressed animals to maintain a response to a new challenge. In this sense the sensitization serves an adaptive purpose. However, the finding that LC neurons of repeatedly stressed animals became responsive to subthreshold levels of CRH (and presumably subthreshold challenges that activate the LC via CRH) suggests that the consequences of LC activation (i.e., increased arousal, loss of focused attention, changes in colon motility, and decreases in immune function) could be elicited inappropriately and may persist. Thus, LC sensitization to CRH in animals with a history of repeated or chronic stress could play a role in some of these symptoms of depression or other stress-related psychiatric disorders, such as posttraumatic stress disorder. Given the proposed role of the LC in CRH-induced colonic motility and its responsivity to colon distention, this phenomenon may also be important in stress-related bowel disorders, such as irritable bowel syndrome. In this disorder, which is also associated with anxiety, mechanisms regulating colonic motility becomes sensitized such that lower levels of distention elicit a motor response. LC sensitization to CRH could play a role in both the motor and cognitive symptoms of this disorder.

The mechanism of LC sensitization to CRH has yet to be elucidated. Receptor binding studies suggest that alterations in CRH receptor binding kinetics do not play a role in LC sensitization (Curtis *et al.,* 1995). It is possible that changes at the level of receptor transduction mechanisms are involved in this sensitization. Alternatively, increased expression of factors that potentiate the postsynaptic effects of CRH (such as vasopressin) could be important in this phenomenon. Interestingly, stress-induced facilitation of the LC–NE system is also apparent at the level of NE release in target regions. Thus, stress-elicited NE release in the hippocampus (measured by microdialysis) is enhanced in animals with a history of chronic stress (Nisenbaum *et al.,* 1991; Nisenbaum and Abercrombie, 1993).

B. LC Activity in Adrenalectomized Animals

Like repeated stress, adrenalectomy increases the neurohormone activity of CRH. This is expressed as increased CRH mRNA expression in paraventricular hypothalamic neurons and increased CRH synthesis and release into the median eminence (Paull and Gibbs, 1983; Plotsky and Sawchenko, 1987; Bradbury *et al.,* 1991; Dallman *et al.,* 1992). These changes are due to removal of corticosteroid feedback inhibition and can be reversed by glucocorticoid replacement. CRH that impacts on the LC is similarly regulated by glucocorticoids. Several findings suggest that tonic and stress-elicited CRH release in the LC is increased in adrenalectomized rats (Pavcovich and Valentino, 1997). Thus, LC discharge rates are higher in adrenalectomized vs. sham-operated rats, and this difference is reversed by intracoerulear administration of CRH antagonists into the LC (Figure 6). Moreover, a comparison of the CRH dose-response curves in adrenalectomized vs. sham rats suggests that CRH receptors are occupied prior to administration of exogenous CRH in adrenalectomized, but not sham, rats. These findings suggest a parallel regulation of the neurohormone CRH that impacts on the pituitary and the neurotransmitter CRH that impacts on the LC. Parallel regulation of these two systems may underlie the coexistence of endocrine and cognitive dysfunctions that occur in depression.

VII. CRH-LC Link in Psychiatric Disorders: Pharmacological Studies

Depression, which has long been associated with stress, is characterized by both neuroendocrine dysfunctions and dysfunctions in biogenic amine systems. Just as the CRH-LC link may integrate neuroendocrine with nonendocrine responses to stress, this link may integrate neuroendocrine, immune, and cognitive symptoms of depression in subjects where CRH-LC interactions are altered, perhaps by

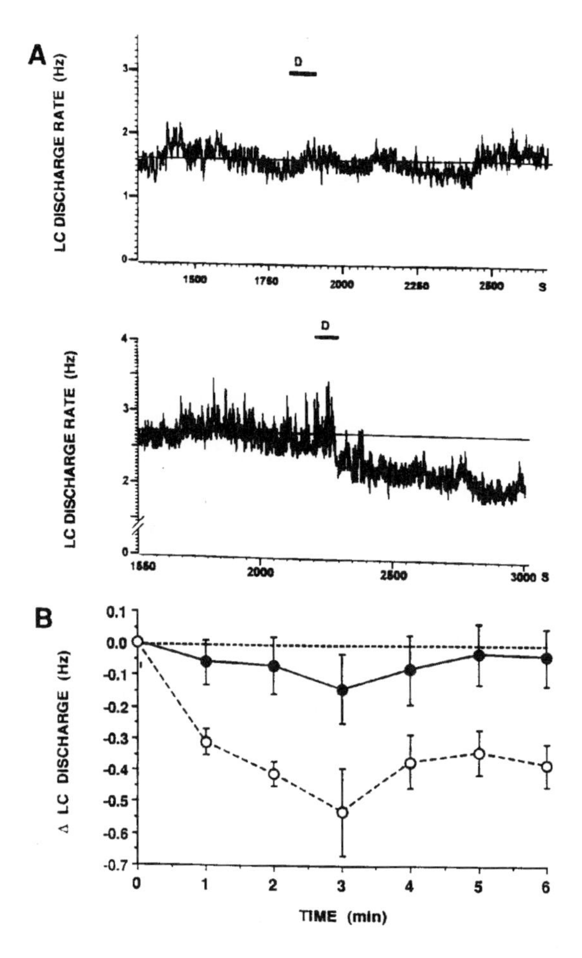

FIGURE 6 Adrenalectomy results in enhanced tonic CRH release in the LC. (A) Continuous chart record of LC discharge rate in a sham (top) or adrenalectomized (bottom) rat before and after intracoerulear infusion of the CRH antagonist, D-Phe CRH_{12-41}. The abscissae indicate time(s). The ordinates indicate LC discharge rate (Hz). The time of infusion is indicated by the bar and *D* above the traces. The solid horizontal line represents the mean LC discharge rate determined over 6 min prior to D-Phe CRF_{12-41} infusion. Top and bottom traces were from single neurons recorded in a sham-operated and adrenalectomized rat, respectively. (B) Mean effect of D-Phe CRH_{12-41} in sham-operated (closed symbols, $n = 8$) vs. adrenalectomized rats (open symbols, $n = 6$). The abscissa indicates time after D-Phe CRH_{12-41} infusion. The ordinate indicates the change in LC discharge rate from baseline (Hz). The effects of D-Phe CRH_{12-41} were significantly different in adrenalectomized vs. sham-operated rats ($F(1,97) = 8.3, p < 0.02$). Additionally, D-Phe CRH_{12-41} significantly decreased LC discharge rate in adrenalectomized ($F(5,41) = 4.9, p < 0.002$), but not sham-operated ($F(7,55) = 0.72$) rats. See Pavcovich and Valentino, 1997 for details. Reprinted from *Journal of Neuroscience*, 17 (1), by L. A. Pavcovich and R. J. Valentino, "Regulation of a putative neurotransmitter effect of cortocotropin-releasing factor: Effects of adrenalectomy," 401–408, 1997, with kind permission of the Society for Neuroscience.

prior stress. The role of the CRH-LC link in depression was investigated in pharmacological studies designed to detemine whether antidepressants interfere with CRH in the LC.

A. Antidepressant Effects on CRH-LC Interactions

The effects of several, pharmacologically distinct antidepressants on LC spontaneous discharge rate, LC responses to sciatic nerve stimulation, and LC activation by endogenous and exogenous CRH were evaluated (Valentino *et al.,* 1990; Curtis and Valentino, 1991; 1994). Chronic (but not acute) administration of the norepinephrine reuptake inhibitor, desmethylimipramine, and the atypical antidepressant, mianserin, prevented LC activation by hypotensive stress. LC spontaneous discharge rate, activation by sciatic nerve stimulation, and activation by exogenously administered CRH were similar in rats chronically administered desmethylimipramine, mianserin, or vehicle. Although the site at which antidepressants act to prevent LC activation by hypotension is unknown, the lack of antidepressant effect on LC responses to exogenous CRH suggests a presynaptic site of action. Thus, chronic administration of these antidepressants could reduce stress-elicited CRH release in the LC, although this could occur at several points proximal to the putative CRH-LC synapse.

In contrast to desmethylimipramine and mianserin, chronic administration of the serotonin reuptake inhibitor, sertraline, or the monoamine oxidase inhibitor, phenelzine, did not alter LC activation by exogenous or endogenous CRH (Valentino *et al.,* 1990; Curtis and Valentino, 1994). However, rats chronically administered these antidepressants exhibited a greater signal-to-noise ratio of the LC sensory response. This effect is opposite to that of CRH, and may serve to oppose the disruptive effects of endogenous CRH on this response. As previously discussed, a decrease in the signal-to-noise ratio of the LC sensory response, as is produced by CRH or stress, may be related to shifts from focused to scanning attention and the possible disruption of behaviors requiring focused attention.

Importantly, these studies demonstrated that four pharmacologically distinct antidepressants share the potential to interfere with the CRH-LC link at some level; either attenuating stress-elicited CRH activation of the LC (desmethylimipramine and mianserin) or producing effects on LC neurons that are opposed to those of CRH (sertraline and phenelzine). Taken together, the results suggest that the CRH-LC link may be one target of antidepressant action.

VIII. Conclusions

Convergent neurochemical and physiological findings obtained over the past two decades have indicated that the LC–NE system is potently activated by stressors. Recent anatomical and electrophysiological studies that address the location and neurochemical identity of afferents to the LC are beginning to elucidate the circuitry and neuromediators of this activation. These studies have implicated excitatory amino acids and CRH as likely mediators of LC activation by stressors, although they do not rule out the role of other mediators or LC afferents. The identity and influence of afferent input to LC dendrites that extend beyond the nuclear zone must be determined. Importantly, ultrastructural studies that can more precisely characterize substrates for communication between LC neurons and their afferents are needed to reveal how this communication takes place at the cellular level.

An equally important issue to address is the consequence of LC activation by stressors. The physiological characteristics of the LC reviewed in this chapter have long implicated this system in arousal, attention, and vigilance. Some of the findings discussed are consistent with these functions. Thus, stress-elicited LC activation results increased arousal and shifts from focused to labile attention, which might be considered cognitive aspects of the stress response. However, more recent findings implicate the LC in other aspects of the stress response, such as alterations in immune function or gastrointestinal activity. These other potential functions of the LC–NE system need to be examined in more detail, and the efferent circuitry of these responses must be elucidated. A direct role of the LC–NE system in these aspects of stress responses would suggest that it has a more global influence on the stress response, perhaps facilitating the integration between certain cognitive, immune, and autonomic responses.

Finally, stress has been linked to variety of medical and psychiatric disorders, which are characterized by some combination of neuroendocrine, immune, autonomic, and cognitive symptoms. Examples of these disorders include depression, posttraumatic stress disorder, and irritable bowel syndrome. The studies reviewed herein implicate the CRH-LC connection in these "integrative" disorders and suggest that drugs that target this link may be useful in ameliorating some of the symptoms of these diseases. Future studies that directly examine the role of the CRH-LC connection in these disorders could lead to potential treatments of these debilating stress-related diseases.

References

Abercrombie, E. D., & Jacobs, B. L. (1987). Single unit response of noradrenergic neurons in locus coeruleus of freely moving cats. I. Acutely presented stressful and nonstressful stimuli. *Journal of Neuroscience* **7,** 2837–2843.

Abercrombie, E. D., Keller, R. W., & Zigmond, M. J. (1988). Characterization of hippocampal norepinephrine release as measured by microdialysis perfusion: Pharmacological and behavioral studies *Neuroscience* **27,** 897–904.

Ader, J. P., Room, P, Postema, F., & Korf, J. (1980). Bilaterally diverging axon collaterals and contralateral projections from rat locus coeruleus neurons, demonstrated by fluorescent retrograde double labeling and norepinephrine metabolism. *Journal of Neural Transmission* **49,** 207–208.

Akaoka, H., & Aston-Jones, G. (1991). Opiate withdrawal-induced hyperactivity of locus coeruleus neurons is substantially mediated by augmented excitatory amino acid input. *Journal of Neuroscience* **11,** 3830–3839.

Anden, N.-E., Dahlstrom, A., Fuxe, K., Larsson, K., Olson, L., & Ungerstedt, U. (1966). Ascending monoamine neurons to the telencephalon and diencephalon. *Acta Physiol. Scand.* **67,** 313–326.

Aston-Jones, G., Astier, B., Ennis, M. (1992). Inhibition of locus coeruleus noradrenergic neurons by C1 adrenergic cells in the rostral ventral medulla. *Neuroscience* **48,** 371–382.

Aston-Jones, G., & Bloom, F. E. (1981a). Activity of norepinephrine-containing locus coeruleus neurons in behaving rats anticipates fluctuations in the sleep-waking cycle. *Journal of Neuroscience* **1,** 876–886.

Aston-Jones G., & Bloom, F. E. (1981b). Norepinephrine-containing locus coeruleus neurons in behaving rats exhibit pronounced responses to non-noxious environmental stimuli. *Journal of Neuroscience* **1,** 887–900.

Aston-Jones, G., Ennis, M., Pieribone, V. A., Nickell, W. T., & Shipley, M. T. (1986). The brain nucleus locus coeruleus: Restricted afferent control of a broad efferent network. *Science* **234,** 734–737.

Aston-Jones, G., Shipley, M. T., Chouvet, G., Ennis, M., Van Bockstaele, E. J., Pieribone, V., Shiekhattar, R., Akaoka, H., Drolet, G., Astier, B., Charlety, P., Valentino, R., & Williams, J. T. (1991). Afferent regulation of locus coeruleus neurons: Anatomy, physiology and pharmacology. *Progress in Brain Research* **85:** 47–75.

Berkenbosch, F., van Oers, J., Del Rey, A., Tilders, F., & Besedovsky, H. (1987). Corticotropin-releasing factor-producing neurons in the rat activated by interleukin-1. *Science,* **238,** 524–526.

Berridge, C. W., & Foote, S. L. (1991). Effects of locus coeruleus activation on electroencephalographic activity in the neocortex and hippocampus. *Journal of Neuroscience* **11,** 3135–3145.

Berridge, C. W., Page, M. E., Valentino, R. J., & Foote, S. L. (1993). Effects of locus coeruleus inactivation on electroencephalographic activity in neocortex and hippocampus. *Neuroscience* **55,** 381–393.

Bradbury, M. J., Akana, S. F., Cascio, C. S., Levin, N., Jacobson, L., & Dallman, M. F. (1991). Regulation of basal ACTH secretion by corticosterone is mediated by both type I (MR) and type II (GR) receptors in rat brain. *Journal of Steroid Biochemistry and Molecular Biology* **40,** 133–142.

Britton, D. R., Koob, G. F., Rivier, J. and Vale, W. (1982). Intraventricular corticotropin-releasing factor enhances behavioral effects of novelty. *Life Science* **31,** 363–367.

Britton, K. T., Lee, G., Vale, W., Rivier, J., & Koob, G. F. (1986). Corticotropin-releasing factor (CRF) receptor antagonist blocks activating and "anxiogenic" actions of CRF in the rat. *Brain Research* **369,** 303–306.

Britton, K., Morgan, J., Rivier, J., Vale, W., & Koob, G. (1985). Chlordiazepoxide attenuates CRF-induced response suppression in the conflict test. *Psychopharmacology* **86,** 170–174.

Britton, K. T., Segal, D. S., Kuczenski, R., & Hauger, R. (1992). Dissociation between in vivo hippocampal norepinephrine response and behavioral/neuroendocrine responses to noise stress in rats. *Brain Research* **574,** 125–130.

Brown, M. R., & Fisher, L. A. (1985). Corticotropin-releasing factor: Effects on the autonomic nervous systems and visceral systems. *Federation Proceedings* **44,** 243–248.

Brown, M. R., Fisher, L. A., Webb, V., Vale, W., & Rivier, J. (1985). Corticotropin-releasing factor: A physiologic regulator of adrenal epinephrine secretion. *Brain Research* **328,** 355–357.

Brown, D. L., & Guyenet, P. G. (1986). Cardiovascular neurons of brainstem with projections to spinal cord. *American Journal of Physiology* **247,** R1009–R1026.

Brun, P., Suaud-Chagny, M. F., Gonon, F., & Buda, M. (1993). In vivo noradrenaline release evoked in the anteroventral thalamic nucleus by locus coeruleus activation: An electrochemical study. *Neuroscience* **52,** 961–972.

Brun, P., Suaud-Chagny, M. F., Lachuer, J., Gonon, F., & Buda, M. (1991). Catecholamine metabolism in locus coeruleus neurons: A study of its activation by sciatic nerve stimulation in the rat. *European Journal of Neuroscience* **3,** 397–406.

Butler, P. D., Weiss, J. M, Stout, J. C., & Nemeroff, C. B. (1990). Corticotropin-releasing factor produces fear-enhancing and behavioral activating effects following infusion into the locus coeruleus. *Journal of Neuroscience* **10,** 176–183.

Cassens, G., Roffman, G., Kuruc, A., Orsulak, P. J., & Schildkraut, J. J. (1980). Alterations in brain norepinephrine metabolism induced by environmental stimuli previously paired with inescapable shock. *Science* **209,** 1138–1139.

Chen, M. F., Chiu, T. H., & Lee, E. H. Y. (1992). Noradrenergic mediation of the memory-enhancing efect of corticotropin-releasing factor in the locus coeruleus. *Psychoneuroendocrinology* **17,** 113–124.

Chiang, C., & Aston-Jones, G. (1993). Response of LC neurons to footshock stimulation is mediated by neurons in the rostral ventral medulla. *Neuroscience* **53,** 705–715.

Chiang, C., Curtis, A. L., Drolet, G., Valentino, R. J., & Aston-Jones, G. (1991). Auditory responses of locus coeruleus neurons are attenuated by excitatory amino acid receptor antagonists in the awake rat. *Society for Neuroscience Abstracts* **17,** 1540.

Chuluyan, G., Saphier, D., Rohn, W., & Dunn, A. J. (1992). Noradrenergic innervation of the hypothalamus participates in the adrenocortical responses to interleukin-1. *Neuroendocrinology* **56,** 106–111.

Cole, B. J., & Koob, G. F. (1988). Propranolol antagonizes the enhanced condition fear produced by corticotropin-releasing factor. *Journal of Pharmacology and Experimental Therapeutics* **247,** 902–910.

Cummings, S., Elde, R., Ells, J., & Lindall, A. (1983). Corticotropin-releasing factor immunoreactivity is widely distributed within the central nervous system of the rat: An immunohistochemical study. *Journal of Neuroscience* **3,** 1355–1368.

Cunnick, J. E., Lysle, D. T., Kucinski, B. J., & Rabin, B. S. (1990). Evidence that shock-induced immune suppression is mediated by adrenal hormones and peripheral B-adrenergic receptors. *Pharmcology Biochemistry and Behavior* **36,** 645–651.

Cunningham, E. T. J., & Sawchenko, P. E. (1988). Anatomical specificity of noradrenergic inputs to the paraventricular and supraoptic nuclei of the rat hypothalamus. *Journal of Comparative Neurology* **274,** 60–76.

Curtis, A. L., Drolet, G., & Valentino, R. J. (1993). Hemodynamic stress activates locus coeruleus neurons of unanesthetized rats. *Brain Research Bulletin* **31,** 737–744.

Curtis, A. L., Grigoradis, D., Page, M. E., Rivier, J., & Valentino, R. J. (1994). Pharmacological comparison of two corticotropin-releasing factor antagonists: In vivo and in vitro studies. *Journal of Pharmacology and Experimental Therapeutics* **268,** 359–365.

Curtis, A. L., Grigoriadis, D., Pavcovich, L. A., & Valentino, R. J. (1995). Prior stress alters corticotropin-releasing factor neurotransmission in the locus coeruleus. *Neuroscience* **65,** 541–550.

Curtis, A. L., Lechner, S. M., Pavcovich, L. A., & Valentino, R. J. (1997). Activation of the locus coeruleus noradrenergic system by intracoerulear microinfusion of corticotropin-releasing factor: Effects on discharge rate, cortical norepinephrine levels and cortical electroencephalographic activity. *Journal of Pharmacology and Experimental Therapeutics* ***281,*** 163–172.

Curtis, A. L., & Valentino, R. J. (1991). Acute and chronic effects of the atypical antidepressant, mianserin, on brain noradrenergic neurons. *Psychopharmacology* **103,** 330–338.

Curtis, A. L., & Valentino, R. J. (1994). Corticotropin-releasing factor neurotransmission in locus coeruleus: A possible site of antidepressant action. *Brain Research Bulletin* **35,** 581–587.

Dallman, M. F., Akana, S. F., Scriber, K. A., Bradbury, M. J., Walker, C.-D., Strack, A. M., & Cascio, C. S. (1992). Stress, feedback and facilitation in the hypothalamo-pituitary-adrenal axis. *Journal of Neuroendocrinology* **4,** 517–526.

Dietrichs, E. (1985). Divergent axon collaterals to cerebellum and amygdala from neurons in the parabrachial nucleus, the nucleus locus coeruleus and some adjacent nuclei. A fluorescent double labeling study using rhodamine labeled latex microspheres and fast blue as retrograde tracers. *Anatomical Embryology* **172,** 375–382.

Drolet, G., & Aston-Jones, G. (1991). Putative glutamatergic afferents to the nucleus locus coeruleus from the nucleus paragigantocellularis: Immunohistochemistry and tract-tracing. *Society for Neuroscience Abstracts* **17,** 1541.

Drolet, G., Van Bockstaele, E. J., & Aston-Jones, G. (1992). Robust enkephalin innervation of the locus coeruleus from the rostral medulla. *Journal of Neuroscience* **123,** 3162–3174.

Dunn, A. J. (1992). Endotoxin-induced activation of cerebral catecholamine and serotonin metabolism-comparison with interleukin-1. *Journal of Pharmacology and Experimental Therapeutics* **261,** 964–969.

Dunn, A. J., & Berridge, C. W. (1987). Corticotropin-releasing factor administration elicits a stress-like activation of cerebral catecholaminergic systems. *Pharmacology Biochemistry and Behavior* **27,** 685–691.

Eaves, M., Thatcher-Britton, K., Rivier, J., Vale, W., & Koob, G. (1985). Effects of corticotropin-releasing factor on locomotor activity in hypophysectomized rats. *Peptides* **6,** 923–926.

Ennis, M., & Aston-Jones, G. (1988). Activation of locus coeruleus from nucleus paragigantocellularis: A new excitatory amino acid pathway in brain. *Journal of Neuroscience* **8,** 3644–3657.

Ennis, M., & Aston-Jones, G. (1989a). Potent inhibitory input to locus coeruleus from the nucleus prepositus hypoglossi. *Brain Research Bulletin* **22,** 793–803.

Ennis, M., & Aston-Jones, G. (1989b). GABA-mediated inhibition of locus coeruleus from the dorsomedial rostral medulla. *Journal of Neuroscience* **9,** 2973–2981.

Ennis, M., Behbenani, M. M., Van Bockstaele, E. J., Shipley, M. T., & Aston-Jones, G. (1991). Projections from the periaqueductal gray to the rostromedial pericoerulear region and nucleus locus coeruleus: Antaomic and physiologic studies. *Journal of Comparative Neurology* **306,** 480–494.

Fisher, L. A., Rivier, J., Rivier, C., Spiess, J., Vale, W. W., & Brown, M. R. (1982). Corticotropin-releasing factor (CRF): Central effects on mean arterial pressure and heart rate in rats. *Endocrinology* **110,** 2222–2224.

Florin-Lechner, S. M., Druhan, J., Aston-Jones, G., & Valentino, R. J. (1996). Enhanced norepinephrine release in prefrontal cortex with burst stimulation of the locus coeruleus. *Brain Research* **742,** 89–97.

Foote, S. L., G. Aston-Jones, & Bloom, F. E. (1980). Impulse activity of locus coeruleus neurons in awake rats and monkeys is a function of sensory stimulation and arousal. *Proceedings of the National Academy of Science (USA)* **77,** 3033–3037.

Foote, S. L., Aston-Jones, G., & Bloom, F. E. (1983). Nucleus locus coeruleus: New evidence of anatomical and physiological specificity. *Physiological Reviews* **63,** 844–914.

Foote, S. L., Freedman, R., & Oliver, A. P. (1975). Effects of putative neurotransmitter on neuronal activity in monkey auditory cortex. *Brain Research* **86,** 229–242.

Forloni, G., Grzanna, R., Blakely, R. D., & Coyle, J. T. (1987). Co-localization of N-acetyl-aspartyl-glutamate in central cholinergic, noradrenergic, and serotonergic neurons. *Synapse* **1,** 455–460.

Freedman, R., Foote, S. L., & Bloom, F. E. (1975). Histochemical characterization of a neocortical projection of the nucleus locus coeruleus in the squirrel monkey. *Journal of Comparative Neurology* **164,** 209–232.

Freedman, R., Hoffer, B. J., Woodward, D. J., & Puro, D. (1977). Interaction of norepinephrine with cerebellar activity evoked by mossy and climbing fibers. *Experimental Neurology* **55,** 269–288.

Fritschy, J.-M., & Grzanna, R. (1990). Distribution of locus coeruleus axons within the rat brainstem demonstrated by PHA-L anterograde tracing in combination with dopamine-B-hydroxylase immunofluorescence. *Journal of Comparative Neurology* **293,** 616–631.

Fritschy, J.-M., Lyons, W. E., Mullen, C. A., Kosofsky, B. E., Molliver, M. E., & Grzanna, R. (1987). Distribution of locus coeruleus axons in the rat spinal cord: A combined anterograde transport and immunohistochemical study. *Brain Research* **437,** 176–180.

Grzanna, R., and Molliver, M. E. (1980). The locus coeruleus in the rat: An immunohistochemical delineation. *Neuroscience* **5,** 21–40.

Guyenet, P. G. (1980). The coerulospinal noradrenergic neurons: Anatomical and electrophysiological studies in rat. *Brain Research* **189,** 121–133.

Guyenet, P. G., & Young, B. S. (1987). Projections of nucleus paragigantocellularis lateralis to locus coeruleus and other structures in rat. *Brain Research* **406,** 171–184.

Heureux, R. L., Dennis, T., Curet, O., & Scatton B. (1986). Measurement of endogenous noradrenaline release in the rat cerebral cortex in vivo by transcortical dialysis: Effects of drugs affecting noradrenergic transmission. *Journal of Neurochemistry* **46,** 1794–1801.

Hida, T., & Shimazu, N. (1982). The interrelation between the laterdorsal tegmental area and lumbosacral segments of rats as studied by HRP method. *Archives of Histology, Japan* **45,** 495–504.

Hoffer, B. J., Siggins, G. R., & Bloom, F. E. (1971). Studies on norepinephrine-containing afferents to Purkinje cells of rat cerebellum. II. Sensitivity of Purkinje cells to norepinephrine and related substances administered by microiontophoresis. *Brain Research* **25,** 522–534.

Hoffer, B. J., Siggins, G. R., Oliver, A. P., & Bloom, F. E. (1973). Activation of the pathway from locus coeruleus to rat cerebella Purkinje neurons: Pharmacological evidence of noradrenergic central inhbition. *Journal of Pharmacology and Experimental Therapeutics* **284,** 553–569.

Holmes, P., and Crawley, J. (1995). Coexisting neurotransmitters in central noradrenergic neurons. In F. E. Bloom & D. J. Kupfer (Eds.), *Psychopharmacology: The fourth generation of progress* (pp. 347–354). New York: Raven Press.

Irwin, M. (1993). Stress-induced supression. Role of the autonomic nervous system. *Annals of the New York Academy of Science* **697,** 203–218.

Irwin, M., Hauger, R. L., Jones, L., Provencio, M., & Britton, K. T. (1990). Sympathetic nervous system mediates central corticotropin-releasing factor induced-suppression of natural killer cytotoxicity. *Journal of Pharmacology and Experimental Therapeutics* **255,** 101–107.

Jacobs, B. L. (1987). Central monoaminergic neurons: Single-unit studies in behaving animals. In H. Y. Meltzer (Ed.), *Psychopharmacology: The third generation of progress* (pp. 159–170). New York: Raven Press.

Kalin, N. H. (1985). Behavioral effects of ovine corticotropin-releasing factor administered to rhesus monkeys. *Federation Proceedings* **44,** 249–254.

Kalin, N. H., Sherman, J. E., & Takahashi, L. K. (1988). Antagonism of endogenous corticotropin-releasing hormone systems attenuates stress-induced freezing behaviors in rats. *Brain Research* **457,** 130–135.

Koda, L. Y., Schulman, J. A., & Bloom, F. E. (1978). Ultrastructural identification of noradrenergic terminals in the rat hippocampus; unilateral destruction of the locus coeruleus with 6-hydroxydopamine. *Brain Research* **145,** 190–195.

Korf J., Aghajanian, G. K., & Roth, R. (1973). Increased turnover of norepinephrine in the rat cerebral cortex during stress: Role of the locus coeruleus. *Neuropharmacology* **12,** 933–938.

Krynjevic, K., & Phillis, J. W. (1963). Actions of certain amines on cerebral cortical neurones. *British Journal of Pharmacology* **20,** 471–490.

Lavicky, J., & Dunn, A. J. (1993). Corticotropin-releasing factor stimulates catecholamine release in hypothalamus and prefrontal cortex in freely moving rats as assessed by microdialysis. *Journal of Neurochemistry* **60,** 602–612.

Lechner, S. M., Curtis, A. L., Brons, R. & Valentino, R. J. (1997). Locus coeruleus activation by colon distention: Role of corticotropin-releasing factor and excitatory amino acids. *Brain Research.* **756,** 114–124.

Lenz, H. J., Raedler, A., Geten, H., Vale, W., & Rivier, J. (1988). Stress-induced gastrointestinal secretory and motor responses in rats are mediated by endogenous corticotropin-releasing factor. *Gastroenterology* **95,** 1510–1517.

Loewy, A. D., Saper, C. B., & Baker, R. P. (1979). Descending projections from the pontine micturition center. *Brain Research* **172,** 533–538.

Loughlin, S., Foote, S. L., & Bloom, F. E. (1986a). Efferent projections of nucleus locus coeruleus: Topographic organization of cells of origin demonstrated by 3-D reconstruction. *Neuroscience* **18,** 291–306.

Loughlin, S. E., Foote, S. L., & Grzanna, R. (1986b). Efferent projections of nucleus locus coeruleus: Morphologic subpopulations have different efferent targets. *Neuroscience* **18,** 307–319.

Loy, R., Koziell, D. A., Lindsey, J. D., & Moore, R. Y. (1980). Noradrenergic innervation of the adult rat hippocampal formation. *Journal of Comparative Neurology* **189,** 699–710.

Luppi, P.-H., Aston-Jones, G., Akaoka, H., Chouvet, G. T., & Jouvet, M. (1995). Afferent projections to the rat locus coeruleus demonstrated by retrograde and anterograde tracing with cholera-toxin B subunit and phaseolus vulgaris leucoagglutinin. *Neuroscience* **65,** 119–160.

Marrocco, R. T., Lane, R. F., McClurkin, J. W., Blaha, C. D., & Alkire, M. F. (1987). Release of cortical catecholamines by visual stimulation requires activity in thalmocortical afferents of monkey and cat. *Journal of Neuroscience* **7,** 2756–2767.

McCormick, D. A., & Prince, D. A. (1988). Noradrenergic modulation of firing pattern in guinea pig and cat thalamic neurons, *in vitro. Journal of Neurophysiology* **59,** 978–996.

Melia, K. R., & Duman, R. S. (1991). Involvement of corticotropin-releasing factor in chronic stress regulation of the brain noradrenergic system. *Proceedings of the Natational Academy of Science (U.S.A.)* **88,** 8382–8386.

Melia, K. R., Rasmussen, K., Terwilliger, R. Z., Haycock, J. W., Nestler, E. J., & Duman, R. S. (1992). Coordinate regulation of cyclic AMP system with firing rate and expression of tyrosine hydroxylase in the rat locus coeruleus: Effects of chronic stress and drug treatments. *Journal of Neurochemistry* **58,** 494–502.

Milner, T. A., Morrison, S. F., Abate, C., & Reis, D. J. (1988). Phenylethanolamine N-methyltransferase-containing terminals synapse directly on sympathetic preganglionic neurons in the rat. *Brain Research* **448,** 205–222.

Miyahara, S., & OOmura, Y. (1982). Inhibitor action of the ventral noradrenergic bundle on the lateral hypothalamic neurons through alpha noradrenergic mechanisms in the rat. *Brain Research* **234,** 459–463.

Miyawaki, T., Kawamura, H., Komatsu, K., & Yasugi, T. (1991). Chemical stimulation of the locus coeruleus; inhiitory effects on hemodynamics and renal sympathetic nerve activity. *Brain Research* **568,** 101–108.

Moises, H. C., Burne, R. A. & Woodward, D. J. (1990). Modification of the visual response properties of cerebellar neurons by norepinephrine. *Brain Research* **514,** 259–275.

Moises, H, C., Waterhouse, B. D., & Woodward, D. J. (1981). Locus coeruleus stimulation potentiates Purkinje cell responses to afferent input: The climbing fiber system. *Brain Research* **222,** 42–64.

Moises, H. C., & Woodward, D. J. (1980). Potentiation of GABA inhibitory action in cerebellum by locus coeruleus stimulation. *Brain Research* **182,** 327–344.

Moises, H. C., Woodward, D. J., Hoffer, B. J., & Freedman, R. (1979). Interactions of norepinephrine with Purkinje cell responses to putative amino acid neurotransmitters applied by microiontophoresis. *Experiemental Neurology* **64,** 493–515.

Monnikes, H., Schmidt, B. G., Raybould, H. E., and Tache, Y. (1992a). CRF in the paraventricular nucleus mediates gastric and colonic motor response to restraint stress. *American Journal of Physiology* **262,** G137–G143.

Monnikes, H., Schmidt, B. G., & Tache, Y. (1992b). Corticotropin-releasing factor (CRF) microinfused into the locus ceruleus complex (LCC) stimulates colonic transit in the conscious rats. *Gastroenterology* **102,** A488.

Morilak, D. A., Fornal, C., & Jacobs, B. L. (1987a). Effects of physiological manipulations on locus coeruleus neuronal activity in freely moving cats. II. Glucoregulatory challenge. *Brain Research* **422,** 24–31.

Morilak, D. A., Fornal, C., & Jacobs, B. L. (1987b). Effects of physiological manipulations on locus coeruleus neuronal activity in freely moving cats. II. Thermoregulatory challenge. *Brain Research* **422,** 17–23.

Morrison, J., & Foote, S. (1986). Noradrenergic and serotonergic innervation of cortical, thalamic and tectal visual strutures in old and new world monkeys. *Journal of Comparative Neurology* **243,** 117–128.

Mugnaini, E., & Oertel, W. H. (1985). An atlas of the distribution of GABAergic neurons and terminals in the rat CNS as revealed by GAD immunohistochemistry. In A. Bjorklund & T. Hokfelt (Eds.), *Handbook of Chemical Neuroanatomy.* Vol. 4, GABA and Neuropeptides in the CNS (pp. 436–608). Amsterdam: Elsevier Science Publishers B.V.

Nagai, T., Satoh, K., Imamoto, K., & Maeda, T. (1981). Divergent projections of catecholamine neurons of the locus coeruleus as revealed by fluorescent retrograde double labeling technique. *Neuroscience Letters* **23,** 117–123.

Nakai, Y., & Takaori, S. (1974). Influence of norepinephrine-containing neurons derived from the locus coeruleus on lateral geniculate neuronal activities of cats. *Brain Research* **71,** 47–60.

Nisenbaum, L. K., & Abercrombie, E. D. (1993). Presynaptic alterations associated with enhancement of evoked release and synthesis of norepinephrine in hippocampus of chronically cold stressed rats. *Brain Research* **608,** 280–287.

Nisenbaum, L. K., Zigmond, M. J., Sved, A. F., & Abercrombie, E. D. (1991). Prior exposure to chronic stress results in enhanced synthesis and release of hippocampal norepinephrine in response to a noval stressor. *Journal of Neuroscience* **11,** 1478–1484.

Olson, L., & Fuxe, K. (1971). On the projections from the locus coeruleus noradrenaline neurons: The cerebellar connection. *Brain Research* **28,** 165–171.

Page, M. E., & Abercrobie, E. D. (1995). Corticotropin-releasing factor in the locus coeruleus increases extracellular norepinephrine in the rat hippocampus. *Society of Neuroscience Abstract* **21,** 634.

Page, M. E., Akaoka, H., Aston-Jones, G. & Valentino, R. J. (1992). Bladder distention activates locus coeruleus neurons by an excitatory amino acid mechanism. *Neuroscience* **51,** 555–563.

Page, M. E., Berridge, C. W., Foote, S. L., and Valentino, R. J. (1993). Corticotropin-releasing factor in the locus coeruleus mediates EEG activation associated with hypotensive stress. *Neuroscience Letters* **164,** 81–84.

Paull, W. K., & Gibbs, F. P. (1983). The corticotropin-releasing factor (CRF) neurosecretory system in intact, adrenalectomized, and adrenalectomized-dexamethasone treated rats. *Histochemistry* **78,** 303–316.

Pavcovich, L. A., & Valentino, R. J. (1997). Regulation of a putative neurotransmitter effect of corticotropin-releasing factor: Effects of Adrenalectomy. *Journal of Neuroscience* **17,** 401–408.

Phillis, H. W., Tebecis, A. K., & York, D. H. (1967). The inhibitory action of monoamines on lateral geniculate neurones. *Journal of Physiology (London)* **190,** 563–581.

Pickel, V. M., Joh, T. H., Reis, D. J., Leeman, S. E., & Miller, R. J. (1979). Electron microscopic localization of substance P and enkephalin in axon terminals related to dendrites of catecholaminergic neurons. *Brain Research* **160,** 387–400.

Pieribone, V. A., & Aston-Jones, G. (1991). Adrenergic innervation of the rat nucleus locus coeruleus arises from the C1 and C3 cell groups in the rostral medulla: An anatomic study combining retrograde transport and immunofluorescence. *Neuroscience* **4,** 525–542.

Plotsky, P. M., & Sawchenko, P. E. (1987). Hypophysial-portal plasma levels, median eminence content, and immunohistochemical staining of corticotropin-releasing factor, vasopressin and oxytocin after pharmacological adrenalectomy. *Endocrinology* **120,** 1361–1369.

Potter, E., Sutton, S., Donaldson, C., Chen, R., Perrin, M., Lewis, K., Sawchenko, P. E., & Vale, W. (1994). Distribution of corticotropin-releasing factor receptor mRNA expression in the rat brain and pituitary. *Proceedings of the National Academy of Science (U.S.A.)* **91,** 8777–8781.

Rajkowski, J., Kubiak, P., & Aston-Jones, G. (1992). Activity of locus coeruleus neurons in behaving monkeys varies with focused attention: Short- and long-term changes. *Society for Neuroscience Abstracts* **18:** 538.

Rajkowski, J., Kubiak, P., & Aston-Jones, G. (1994). Activity of locus coeruleus neurons in monkey: Phasic and tonic changes correspond to altered vigilance. *Brain Research Bulletin* **35,** 607–616.

Rasmussen, K., & Aghajanian, G. K. (1989). Withdrawal-induced activation of locus coeruleus neurons in opiate-dependent rats: Attenuation by lesion of the nucleus paragigantocellularis. *Brain Research* **505,** 346–350.

Rasmussen, K., Beitner-Johnson, D. B., Krystal, J. G., Aghajanian, G. K., & Nestler, E. J. (1990). Opiate withdrawal and rat locus coeruleus: Behavioral, electrophysiological and biochemical correlates. *Journal of Neuroscience* **10:** 2308–2317.

Rassnick, S., Sved, A. F., & Rabin, B. S. (1994). Locus coeruleus stimulation by corticotropin-releasing hormone suppresses in vitro cellular immune respones. *Journal of Neuroscience* **14,** 6033–6040.

Rizvi, T. A., Ennis, M. Aston-Jones, G., Jiang, M., Liu, W.-L., Behbehani, M. M., & Shipley, M. T. (1994). Preoptic projections to Barrington's nucleus and the pericoeroulear region: Architecture and terminal organization *Journal of Comparative Neurology* **347,** 1–24.

Room, P., Postema, F., & Korf, J. (1981). Divergent axon collaterals of rat locus coeruleus neurons demonstrated by a fluorescent double labeling technique. *Brain Research* **221,** 219–230.

Ross, C. A., Armstrong, D. A., Ruggiero, D. A., Pickel, V. M., John, T. H., & Reis, D. J. (1981). Adrenaline neurons in the rostral ventrolateral medulla innervate thoracic spinal cord: A combined immunocytochemical and retrograde transport demonstration. *Neuroscience Letters* **25,** 257–262.

Sakanaka, M., Shibasaki, T., & Lederes, K. (1987). Corticotropin-releasing factor-like immunoreactivity in the rat brain as revealed by a modified cobalt-glucose oxide-diaminobenzidene method. *Journal of Comparative Neurology* **260,** 256–298.

Sapolsky, R., Rivier, C., Yamamooto, G., Plotsky, P. & Vale, W. (1987). Interleukin-1 stimulates the secretion of hypothalamic corticotropin-releasing factor. *Science* **238,** 522–524.

Sawchenko, P. E., & Swanson, L. W. (1982). The organization of noradrenergic pathways from the brainstem to the paraventricular and supraoptic nuclei in the rat. *Brain Research Reviews* **4,** 275–325.

Segal, M., & Bloom, F. E. (1974a). The action of norepinephrine in the rat hippocampus. I. Iontophoretic studies. *Brain Research* **72,** 79–97.

Segal, M., & Bloom, F. E. (1974b). The action of norepinephrine in the rat hippocampus. II. Activation of the input pathway. *Brain Research* **72,** 99–114.

Shipley, M. T., Fu, L., Ennis, M., Liu, W. & Aston-Jones, G. (1996). Dendrites of locus coeruleus neurons extend preferentially into two pericoerulear zones. *Journal of Comparative Neurology* **365,** 56–68.

Shipley, M. T., Halloran, F. J., & De La Torre, J. (1985). Surprisingly rich projection from locus coeruleus to the olfactory bulb in the rat. *Brain Research* **329,** 294–299.

Smagin, G. N., Swiergiel, A. H., & Dunn, A. J. (1994). Sodium nitroprusside infusions activate cortical and jypothalamic noradrenergic systems in rats. *Neuroscience Research Communications* **14,** 85–91.

Smagin, G. N., Swiergiel, A. H., & Dunn, A. J. (1995). Corticotropin-releasing factor administered into the locus coeruleus, but not the parabrachial nucleus, stimulates norepinephrine release in the prefrontal cortex. *Brain Research Bulletin* **36,** 71–76.

Steindler, D. A. (1981). Locus coeruleus neurons have axons that branch to the forebrain and cerebellum. *Brain Research* **223,** 367–373.

Stephens, R. L., Jr., Yang, H., Rivier, J., & Tache, Y. (1988). Intracisternal injection of CRF antagonist blocks surgical stress-induced inhbition of gastric secretionin the rat. *Peptides* **9,** 1067–1070.

Strausbaugh, H., & Irwin, M. (1992). Central corticotropin releasing hormone reduces cellular immunity. *Brain Behavior Immunology,* **6:** 11–17.

Sutton, R. E., Koob, G. F., LeMoal, M., Rivier, J., & Vale, W. (1982). Corticotropin-releasing factor produces behavioral activation in rats. *Nature* **297,** 331–333.

Sved, A. F., & Felsten, G. (1987). Stimulation of the locus coeruleus decrease arterial pressure. *Brain Research* **414,** 119–132.

Svensson, T. H. (1987). Peripheral, autonomic regulation of locus coeruleus noradrenergic neurons in brain. Putative implications for psychiatry and psychopharmacology. *Psychopharmacology* **92,** 1–7.

Swanson, L. W. (1976). The locus coeruleus: A cytoarchitectonic, golgi and immunohistochemical study in the albino rat. *Brain Research* **110,** 39–56.

Swanson, L. W., & Hartman, B. K. (1976) The central adrenergic system. An immunofluorescence study of the location of cell bodies and their efferent connections in the rat using dopamine-B-hydroxylase as a marker. *Journal of Comparative Neurology* **163,** 467–506.

Swanson, L. W., Sawchenko, P. E., Rivier, J., & Vale, W. W. (1983). Organization of ovine corticotropin-releasing factor immunoreactive cells and fibers in the rat brain: An immunohistochemical study. *Neuroendocrinology* **36,** 165–186.

Swiergel, A. H., Takahashi, L. K., Ruben, W. W., & Kalin, N. H. (1992). Antagonism of corticotropin-releasing factor receptors in the locus coeruleus attenuates shock-induced freezing in rats. *Brain Research* **587,** 263–268.

Tache, Y., Goto, Y., Gunion, M., Vale, W., Rivier, J., & Brown, M. (1983). Inhibition of gastric acid secretion in rats by intracerebral injection of corticotropin-releasing factor (CRF). *Science* **222,** 935–937.

Tache, Y., & Gunion, M. (1985). Corticotropin-releasing factor: Central action to influence gastric secretion. *Federation Proceedings* **44,** 255–258.

Tache, Y., Monnikes, H., Bonaz, B., & Rivier, J. (1993). Role of CRF in stress-related alterations of gastric and colonic motor function. *Annals of the New York Academy of Science* **697,** 233–243.

Tazi, A., Dantzer, E., LeMoal, M., Rivier, J., Vale, W., & Koob, G. F. (1987). Corticotropin-releasing factor antagonist blocks stress-induced fighting in rats. *Regulatory Peptides* **18,** 37–42.

Thierry, A.-M., Javoy, F., Glowinski, J., & Kety, S. S. (1968). Effects of stress on the metabolism of norepinephrine, dopamine and serotonin in the central nervous system of the rat: Modification of norepinephrine turnover. *Journal of Pharmacology and Experimental Therapeutics* **163,** 163–171.

Toomin, C., Petrusz, P., & McCarthy, K. (1987). Distribution of glutamate- and aspartic-like immunoreactivity in the rat central nervous system. *Society for Neuroscience Abstracts* **13,** 1562.

Uehara, A., Gottschall, P. E., Dahl, R. R., & Arimura, A. (1987). Interleukin-1 stimulates ACTH release by an indirect action which requires endogenous corticotropin releasing factor. *Endocrinology* **121,** 1580–1582.

Ungerstedt, U. (1971). Stereotaxic mapping of the monoamine pathways in the rat brain. *Acta Physiol. Scand. Suppl.* **367,** 1–48.

Vale, W., Speiss, J., Rivier, C., & Rivier, J. (1981). Characterization of a 41-residue ovine hypothalamic peptide that stimulates secretion of corticotropin and b-endorphin. *Science* **213,** 1394–1397.

Valentino, R. J., & Aston-Jones, G. (1995). Physiological and anatomical determinants of locus coeruleus discharge. Behavioral and clinical implications. In F. E. Bloom & D. J. Kupfer (Eds.), *Psychopharmacology: The fourth generation of progress* (pp. 373–386). New York: Raven Press.

Valentino, R. J., & Aulisi, E. (1987). Carbachol-induced increases in locus coeruleus spontaneous activity are associated with disruption of sensory responses. *Neuroscience Letters* **74,** 297–303.

Valentino, R. J., Chen, S., Zhu, Y., & Aston-Jones, G. (1996). Evidence for divergent projections of corticotropin-releasing hormone neurons of Barrington's nucleus to the locus coeruleus and spinal cord. *Brain Research* **732,** 1–15.

Valentino, R. J., Curtis, A. L., Parris, D., & Wehby, R. G. (1990). Antidepressant actions on brain noradrenergic neurons. *Journal of Pharmacology and Experimental Therapeutics* **253,** 833–840.

Valentino, R. J., de Boer, S., Bicanich, P., Kang, B., & Aston-Jones, G. (1992). Fos-immunoreactivity (F-IR) in brains of rats exposed to inescapable shock or administered corticotropin-releasing factor (CRF). *Society for Neuroscience Abstracts* **18,** 203.

Valentino, R. J., & Foote, S. L. (1987). Corticotropin-releasing factor disrupts sensory responses of brain noradrenergic neurons. *Neuroendocrinology* **45,** 28–36.

Valentino, R. J., & Foote, S. L. (1988). Corticotropin-relesing factor increases tonic but not sensory-evoked activity of noradrenergic locus coeruleus neurons in unanesthetized rats. *Journal of Neuroscience* **8,** 016–1025.

Valentino, R. J., Foote, S. L., & Aston-Jones, G. (1983). Corticotropin-releasing factor activates noradrenergic neurons of the locus coeruleus. *Brain Research* **270,** 363–367.

Valentino, R. J., Foote, S. L., & Page, M. E. (1993). The locus coeruleus as a site for integrating corticotropin-releasing factor and noradrenergic mediation of stress responses. In Y. Tache (Ed.), "Corticotropin-releasing factor and cytokines: Role in the stress response." *Annals of the New York Academy of Science* **697,** 173–188.

Valentino, R. J., Page, M. E., & Curtis, A. L. (1991). Activation of noradrenergic locus coeruleus neurons by hemodynamic stress is due to local release of corticotropin-releasing factor. *Brain Research* **555,** 25–34.

Valentino, R. J., Page, M. E., Van Bockstaele, E., & Aston-Jones, G. (1992). Corticotropin-releasing factor immunoreactive cells and fibers in the locus coeruleus region: Distribution and sources of input. *Neuroscience* **48,** 689–705.

Valentino, R. J., Pavcovich, L. A., & Hirata, H. (1995). Evidence for corticotropin-releasing hormone projections from Barrington's nucleus to the periaqueductal gray region and dorsal motor nucleus of the vagus in the rat. *Journal of Comp. Neurol.* **363,** 402–422.

Valentino, R. J., & Wehby, R. G. (1988). Corticotropin-releasing factor: Evidence for a neurotransmitter role in the locus coeruleus during hemodynamic stress. *Neuroendocrinology* **48,** 674–677.

Van Bockstaele, E. J., Chan, J., & Pickel, V. M. (1996a). Input from central nucleus of the amygdala efferents to pericoerulear dendrites, some of which contain tyrosine hydroxylase immunoreactivity. *Journal of Neuroscience Research* **45,** 289–302.

Van Bockstaele, E. J., Colago, E. E. O., & Valentino, R. J. (1996b). Corticotropin-releasing factor-containing axon terminals synapse onto catecholamine dendrites and may presynaptically modulate other afferents in the rostral pole of the nucleus locus coeruleus in the rat brain. *Journal of Comparative Neurology.* **364,** 523–534.

Van Bockstaele, E. J., Valentino, R. J., & Pickel, V. M. (1996c). Efferent projections of the central nucleus of the amygdala (CNA) to dendrites in the peri-locus coeruleus (LC) area. *Society for Neuroscience, (abstracts)* **22,** 2048.

Wan, W., Vriend, C. Y., Wetmore, L., Gartner, J. G., Greenberg, A. H., & Nance, D. M. (1993). The effects of stress on splenic immune function are mediated by the splenic nerve. *Brain Research Bulletin* **30,** 101–105.

Waterhouse, B. D., Azizi, S. A., Burne, R. A., & Woodward, D. J. (1990). Modulation of rat cortical area 17 neuronal responses to moving visual stimuli during norepinephrine and serotonin microiontophoresis *Brain Research* **514,** 276–292.

Waterhouse, B. D., Lin, C. S., Burne, R. A., & Woodward, D. J. (1983). The distribution of neocortial projection neurons in the locus coeruleus. *Journal of Comparative Neurology* **217,** 418–431.

Waterhouse, B. D., Moises, H. C., & Woodward, D. J. (1980). Noradrenergic modulation of somatosensory cortical neuronal responses to iontophoretically applied putative neurotransmitters. *Experimental Neurology* **69,** 30–49.

Waterhouse, B. D., Moises, H. C., Yeh, H. H., & Woodward, D. J. (1982). Norepinephrine enhancement of inhibitory synaptic mechanisms in cerebellum and cerebral cortex: Mediation by *B*-adrenergic receptors. *Journal of Pharmacology and Experimental Therapeutics* **221,** 495–506.

Williams, J. T., Egan, T. M., & North, R. A. (1982). Enkephalin opens potassium channels on mammalian central neurons. *Nature* **299,** 74–77.

Wong, M.-L., Licinio, J., Pasternak, K. I., & Gold, P. W. (1994). Localization of corticotropin-releasing hormone (CRH) receptor messenger RNA in adult rat brain by in situ hybridization histochemistry. *Endocrinology* **135,** 2275–2279.

Woodward, D. J., Moises, H. C., Waterhouse, B. D., Hoffer, B. J., & Freedman, R. (1979). Modulatory actions of norepinephrine in the central nervous system. *Federation Proceedings* **38,** 2109–2116.

A Model for the Control of Ingestion—20 Years Later

John D. Davis
E. W. Bourne Behavioral Research Laboratory
New York Hospital–Cornell Medical Center
White Plains, New York 10605

I. Background

In 1975 Barbara Collins, Michael Levine, and I published a control theory model of the meal size of rats ingesting a carbohydrate solution during a 30-min test (Davis *et al.,* 1975). Two years later the model was described again in the *Psychological Review* with considerably more discussion of the assumptions, more details on the mathematical derivation, and with additional supporting data (Davis and Levine, 1977). It has now been more than 20 years since the model first appeared. During that time it has been useful to some of us for clarifying our thinking about the mechanism of the control of ingestion in the rat, for discovering new ways of thinking about this problem, and for discovering some new features of the system that control ingestive behavior. Some of the predictions of the model have held up, and some have not. These verifications and discrepancies have appeared over the years in a variety of publications by myself, myself in collaboration with others, and by a few others. Thus it seemed to be appropriate now to review this literature and to see just how well the model has held up over the past 20 years to see what it has taught us about some of the major variables that control ingestive behavior.

The inspiration for the original model came from a number of sources, the principal ones being the work of Vincent Dethier on the feeding behavior of the blow fly, later summarized in his book *The Hungry Fly* (1976), the books by Douglas Riggs (1970), Fred Toates (1975), and David McFarland (1971) who showed us how to apply control theory to behavioral and physiological problems, and, most important of all, the early work of Curt Richter on the ingestion of carbohydrates by rats. It was his, now classic, paper published in 1940 with Katherine Campbell (Richter and Campbell, 1940) describing the effect of variation in the concentration of carbohydrates on intake that first got me thinking about the problem of what controls the size of a meal and how to go about finding out.

They reported that if rats were offered a choice between water and different concentrations of glucose solutions for 24-hr periods, the rats would show an increasing preference for glucose by ingesting more of it than water as its concentration was increased. At concentrations above about 10%, however, the preference for glucose declined as the concentration was increased still further. They showed that this pattern of increasing and then decreasing preference for glucose

0363-0951/98 $25.00

solutions over water was true for maltose, sucrose, and galactose as well. The intake of those solutions when plotted as a function of concentration has the appearance of an inverted V-shaped function. This function was later given the name "preference aversion" function.

This name is misleading because the term, *aversion,* applied to the declining limb of the function suggests that concentrated solutions are somehow aversive to the rat. This is clearly not the case because rats ingested more of those solutions than they did of water. Nevertheless, the term was coined before a full understanding of the shape of this curve was understood, and so it became the widely accepted name for functions of this type.

Figure 1, redrawn from an early paper by Ernits and Corbit (1973), shows that the phenomenon described by Campbell and Richter could also be obtained when rats were offered the solutions for only 60 min a day, and without requiring a choice between it and water. I have added a dashed line parallel to the horizontal axis in the figure to make a point about something the shape of this curve implies about the control of intake that is often overlooked. Note that with a nonmonotonic curve such as this there are an infinite number of pairs of test solutions that will stimulate intakes that are identical in volume. The dashed line arbitrarily drawn at about 15 ml crosses the preference aversion curve at two points corresponding to two concentrations of the test solution. One lies below the peak of the curve, the other above it. Identical intakes were stimulated by two quite different concentrations of the same test solution.

Theoretically, there are an infinite number of such pairs of solutions that produce equal intakes, one generated by a low concentration and the other by a high concentration test solution. Although these pairs of solutions stimulate equal intake, the reason the rat ingests the same volume of the two solutions must be very different. The two must taste different, and they must have different effects on the gastrointestinal tract. Yet the intakes are the same. This means that relying simply on how much a rat ingests during the test cannot tell us what we need to know, which is what determines how much was ingested. Total amount consumed during a test tells us nothing about the mechanism or the controls that produced that particular outcome. This measure simply reflects the outcome of the effects of a number of variables acting together on the behavior of the animal while it is ingesting fluid.

One solution to this problem of accounting for identical intakes of different concentrations of a carbohydrate is to identify the variables that are thought to control how much of a test solution an animal will ingest during a fixed period of time, make some guesses about how they interact to control the behavior of the animal, and then combine this information, supposition and speculation, into a quantitatively explicit model. Clearly the model will be inadequate because if it were not we would have no need to engage in speculation and could merely assert the truth of the model. There is value in model building, however, because it provides a means by which quantitatively specific predictions about the control of meal size can be made and tested. By this process, and only by this process, can

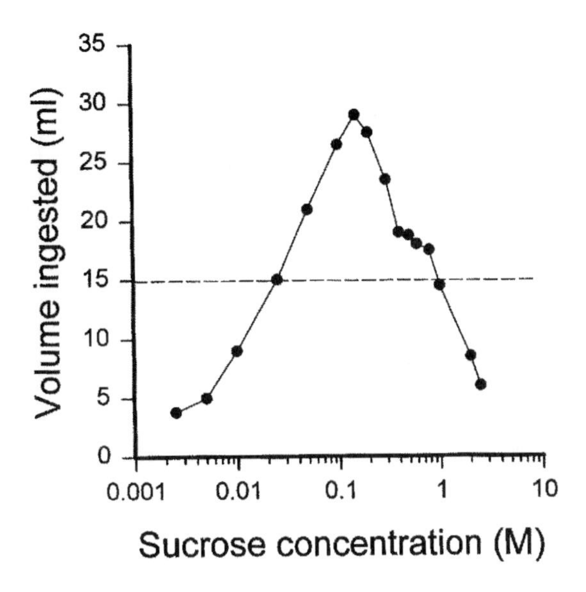

FIGURE 1 Volume of a sucrose solution ingested as a function of concentration. Redrawn from Ernits and Corbit (1973). Copyright © 1973 by the American Psychological Association. Adapted with permission.

we learn where our knowledge about how the system operates is reasonably accurate and where the gaps in our knowledge are.

II. Description of the Model

The model we constructed took into account only two of three major variables known at that time to control intake, namely orosensory stimulation and the postingestional consequences of the ingested fluid. We focused on taste and postingestional consequences of ingestion to the exclusion of food deprivation, and other variables that influence food intake, because it was known by then that these two variables must play a major role in controlling intake. Young and his colleagues (Young, 1967) had shown the importance of gustatory stimulation in controlling intake, and a number of Eliot Stellar's students (Mook, 1963; McClearly, 1953) and others (Shuford, 1959; Jacobs, 1961) had begun to investigate the role of postingestional stimulation in the control of ingestion. We felt we could build on their work to create the outlines of a reasonably good model. Since less was known about the effects of other variables, such as variation in systemic nutrient stores, on the intake of carbohydrate solutions these variables were not treated formally in the model.

The original model is shown in a nonquantitative form in Figure 2. The diagram is organized vertically according to the various anatomical divisions of the body that are involved in the control of ingestion. The multiple lines indicate the flow of fluids through the body. The single lines represent the flow of information

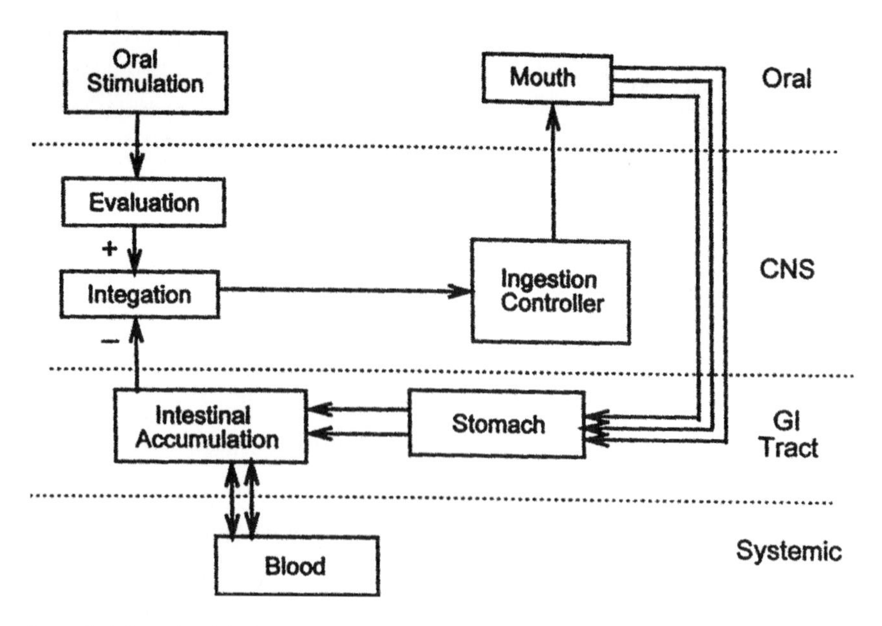

FIGURE 2 Schematic of the Davis and Levine model for the control of meal size in the rat. Redrawn from Davis and Levine (1977). Copyright © 1977 by the American Psychological Association. Adapted by permission.

in the form of neural signals along different pathways from the mouth to the brain and from the gastrointestinal tract to the brain.

For purposes of simplification, and to focus on the variables that were of major interest to us, the various parts of the body involved in feeding were given labels. For example in the upper right of the figure there is a box labeled "Mouth," which is intended to represent all those mechanical components of the ingestive apparatus involved in converting raw food from the environment to a form suitable for processing by the gastrointestinal tract. In this case the teeth, the tongue, the salivary glands, the muscles involved in swallowing, and all of the nerves that control the muscles and glands involved in this activity are lumped together into a single unit because of the common purpose that they serve. This purpose is to start the digestive process, prepare the gastrointestinal tract to receive the incoming load, reduce chunks of food to smaller pieces, mix the food with digestive enzymes, and begin the process of transporting this mouthful through the gastrointestinal tract.

This activity of the mouth is a complicated process in itself, and those interested in understanding the details of it must forgive us for having lumping it all into a single box with a single label. Our interest was in how the properties of test solutions affected the transport of it by the mouth to the gastrointestinal tract rather than in the details of the transport mechanism itself. We wanted to be able to understand how the process of ingestion, once begun, continued and was eventually terminated by the interaction between orophyrangeal and postingestional stimulation. This required identifying and labeling the motor and sensory components of the ingestive apparatus and focusing on them to the exclusion of many important details not immediately relevant to that enterprise.

At the upper left a box labeled "Oral stimulation" represents all the sensory inputs such as the gustatory and somatosensory systems involved in controlling intake once food has been sensed and conveyed to the mouth. I do not recall now why we did not label this box "Oral-olfactory," or "Orophyrangeal stimulation," which would have been better labels for obvious reasons, but I suspect that it was because the role of the gustatory system in the control of intake was better understood than was that of the olfactory system, as is still the case.

The box below this labeled "Evaluation" is included to represent the fact that a particular gustatory stimulus will not necessarily always cause the same reaction to it. Conditioned taste aversions and preferences are the classic examples. A taste that at one time stimulates vigorous intake may at another, because of conditioning, stimulate rejection. Other variables, such as food deprivation, may also operate in whole or in part by altering the evaluation of the flavor of the food. This box is included here to model these facts. The signal entering this box may be amplified, diminished, or reversed in sign depending on the past history of the animal and, perhaps, on its current state of depletion or repletion.

The next level in the model contains an integrator and an ingestion controller. The integrator is the crucial component of the system because its output determines whether ingestion will continue or stop. It is here that information from the tongue and olfactory receptors are combined with information about what is going on in the gastrointestinal tract as the rat ingests the test solution. The signal emerging from the right of this box is the algebraic sum of the inputs to it from gustatory and olfactory stimulation above and from gastrointestinal stimulation entering from below. The magnitude of the signal leaving this box controls the rate of ingestion through its influence on the "Ingestion controller" shown to the right. Thus, whether the animal is eating or not depends on whether this signal is positive or negative. If this signal is greater than zero, its magnitude determines how fast the animal eats. If it is zero or less, then the animal does not eat.

The brain mechanism that translates this signal into the motor activity of the mouth and tongue is modeled by a box labeled "Ingestion controller". It represents the brain mechanisms that are involved in converting the information arriving from the integrator into the motor activity of the mouth and the tongue, the motor activity that we call ingestive behavior. It can be viewed as the final common pathway from the brain to the motor systems involved in transferring food and fluid from the external to the internal environment of the rat.

When the model was originally developed it was assumed that the volume of the intestine was sensed by mechano or stretch receptors and that this information was returned to the "Integration" box as a negative feedback signal. Thus an arrow representing information flow was directed from the box labeled "Intestinal accumulation" to the "Integration" box. This signal was assumed to grow as the gastrointestinal tract filled, and, having a negative sign, subtract from the constant positive signal arriving from the "Evaluation" box. Thus as the intestine fills, the output from the "Integration" box was assumed to diminish in proportion to intestinal volume.

Converting this descriptive model into a quantitatively specific one requires the straightforward application of the mathematics of control theory. For those interested in the process, it is described in detail in our *Psychological Review* article (Davis and Levine, 1977). All that needs to be said here is that the mathematics generates a simple function that describes how ingestion rate changes as the rat continues to eat. This function is the exponential decay function $l(t) = gpde^{-drkt}$, where $l(t)$ is the rate of ingestion at any time, t. In words, this function predicts that a rat will start ingesting the test solution at some rate given by the value of gpd and that this rate will decline over time at a rate given by the value of drk. The individual differences parameter, d, adjusts the evaluated taste of the test solution, gp, and the postingestional negative feedback signal, rk, for the unique characteristics of each rat. It can be seen as a scaling parameter, characteristic of a particular animal, which, although varying from animal to animal, remains constant during a test.

When the function, $l(t)$, is integrated over time it describes the shape of the cumulative intake function, $C(t) = (gp)(rk)^{-1}(1\text{-}e^{-drkt})$. The asymptote of this function, $(gp)(rk)^{-1}$, gives the total amount ingested during the test, and drk measures how fast this asymptote is approached. Thus, the model predicts that the volume ingested in a test will be directly proportional to the stimulating effectiveness of the test solution on the oral cavity and inversely proportional to the magnitude of negative feedback from the gastrointestinal tract generated by the postingestional consequences of ingestion.

When we published this model, we supported it with evidence from the literature and from our work. Since then a considerable body of research has developed that is relevant to an understanding of some of the problems addressed by the model. Therefore, in this review, I have selected from that literature research I feel can contribute to an evaluation of the success of it. It is this literature that I will now review with that objective in mind.

III. Evaluation of the Model

A. Negative Feedback

1. Sham Feeding—Transition from Real Feeding

Probably the most obvious prediction of the model, and the easiest one to evaluate, was that if the negative feedback loop is opened, intake will increase significantly. This follows from the fact that in the absence of a negative feedback signal the input signal will pass through the "Integration" box undiminished, the signal driving the "Ingestion controller" will remain constant, and, therefore, the animal will eat continuously. We knew that something like this would happen because Pavlov had shown many years before that if food eaten by a dog is allowed to flow from an esophageal fistula rather than enter the stomach, the dog will eat continuously for an indefinite period (Pavlov, 1928). This procedure, which has come to be known as sham feeding, was used by Pavlov to collect gastric secretions uncontaminated by ingested food. It was later used by others (Davis and Campbell, 1973; Mook, 1963; Young *et al.,* 1974) to study the effect of the removal of gastrointestinal stimulation on ingestion. Those studies were all consistent in confirming Pavlov's original discovery that when food is not allowed to collect in the gastrointestinal tract the animal will eat much more than normal. The phenomenological prediction was well established.

The prediction made by the model concerning sham feeding, however, was more quantitatively specific and detailed than this. It can be seen in Figure 2 that if the negative feedback signal coming from the gastrointestinal tract is removed the neural information arising from oral stimulation will pass from the mouth through the "Evaluation" and "Integration" boxes unaltered, and the magnitude of the signal reaching the "Ingestion controller" will depend only on this signal, which we assumed would remain constant during the test.

There are two implications of this for the behavior of a rat tested on the sham feeding procedure. One is that once the rat begins ingestion it will continue to do so at a constant rate for as long as the solution is present. Mathematically, with the negative feedback loop open, $l(t) = gpde^{-0t} = gpd$. In other words the rate of ingestion for a rat tested under this condition will depend only on the magnitude of the input signal, g, its evaluation, p, and the individual difference parameter, d. Thus, after the rat first tastes the test solution, it should continue to ingest it at a rate that depends only on the concentration of the solution and the rat's characteristic evaluation of it. Ingestion wil continue indefinitely at a constant rate. Concentrated solutions will be ingested rapidly, and less concentrated solutions more slowly, but there should be no decline in the rate of ingestion during a test.

The other implication, which follows from this, is that if a rat is given successive tests with the sham feeding procedure, its intake will remain constant from one test to the next as long as the duration of the test remains the same. There is nothing in the equation to allow for any variation in the volume of fluid ingested as the rat gains sham feeding experience.

As soon as the implications of the fact that $l(t)$ is constant when negative feedback is absent became apparent, it was clear that that the behavior of the sham feeding rat was at variance with it. The prediction of increased intake with sham feeding was well established, but the two other more specific predictions, (1) that the rate of ingestion during sham feeding will not decline during a test and (2) that intake on sham feeding tests will not change with sham feeding experience, were known to be wrong.

The second of these two implications, that intake on sham feeding tests will remain constant from test to test, could be seen to be wrong from the results of a study by Davis and Campbell (1973). That study introduced a novel surgical procedure for the study of sham feeding in the rat. Sham feeding was accomplished in their procedure by aspirating flud ingested during a test through a tube surgically implanted in the stomach rather than letting it flow from an esophageal fistula in the neck as was done by Pavlov in the dog (1928) and Mook in the rat (1963). With our method, ingested fluids stimulate the oral cavity, the esophagus, and the stomach, rather than just the oral cavity and the upper esophagus. Fluid was not allowed, however, to accumulate in the stomach because it was withdrawn through a surgically implanted gastric catheter as the rat drank.

The finding we reported (Davis and Campbell, 1973), which we later realized was at odds with the model, was that, although intake on the first sham test was larger than on a preceding real feeding test, it continued to increase with further sham feeding testing. Intake reached a maximum only after about four to five consecutive sham feeding tests. We also showed that as the rats gained sham drinking experience, the rate of ingestion during the first 10 min of the test increased. These findings were confirmed soon after by Young and colleagues (1974) using food deprived rats, a different test solution, and a different sham feeding procedure, the gastric fistula, now in common use. Mook and colleagues (1983), using esophagostomized rats, and Weingarten and Kulivosky (1989), using the gastric fistula, later reported similar findings. These reports were all consistent in contradicting the model.

In our report we (Davis and Campbell, 1973) suggested two possible explanations for the progressive increase in intake with sham feeding experience. One was that in changing from real feeding to sham feeding the rat had to adjust to the novelty of not having the stomach and intestine fill as it ate, and that it took four to five days for this adjustment to take place. The other was that there existed some kind of conditioned control over ingestion, which depended on an association between oral stimulation by the test solution and some consequence of its accumulation in the gastrointestinal tract. The progressive rise in intake, according to this interpretation, was the result of the extinction of this conditioned control when oral stimulation from the food was no longer followed by the consequences of gastrointestinal filling.

We favored this latter explanation because of the outcome of one feature of the experiment. This was to test the rats sequentially on two different types of test solutions. Rats were adapted under real feeding conditions to ingest one test solution, and then, when intake was stable, they were given five consecutive sham feeding tests. Following this they were given real feeding tests with another test solution that had a very different flavor. When intake was stable on this new solution, they were given five sham feeding tests with it.

We used diluted sweetened condensed milk as one test solution and Vivasorb, a chemically defined diet, for the other. With both test solutions, and regardless of which test solution was used first, the rats showed the same progressive rise in intake on the second series of sham feeding tests as they had on the first. We took this as evidence against the idea that adaptation to the novelty of sham feeding was the reason the rats took up to five days to achieve maximum intake.

Later Weingarten and Kulikovsky (1989) tested this interpretation using two different procedures. In their first experiment they trained two groups of rats to sham feed a 32% sucrose flavored solution on one series of sham feeding tests with lemon and, on another, with almond. This gave both groups of rats sham feeding experience with both flavors. Following this, members of one of the two groups were given a real feeding experience with one of the two flavors and the other group with the other flavor. On the final test they were again given a series of sham feeding tests. On the first of these sham feeding tests the rats ingesting the flavored sucrose not associated with real feeding in the previous series of real feeding tests ingested significantly more of it than they did of the sucrose solution previously associated with real feeding. Real feeding experience with one of the flavors had reconditioned their inhibitory effect on ingestion. Unfortunately they did not measure the change in the rate of ingestin during the test so it was not possible to determine how the inhibitory effect of prior real feeding experience on ingestion was expressed, but their results, nevertheless, support the extinction hypothesis.

They used the phenomenon of latent inhibition in the second of their two tests of the conditioning hypothesis. Rats that had had sham feeding experience with a uniquely flavored solution and later were given real feeding tests with that solution, followed by another sham feeding test, ingested more on that last sham feeding test than did rats that had not had the initial sham feeding experience with that flavored solution. The sham feeding experience with a uniquely flavored solution prevented conditioning from developing on the following real feeding tests. Latent inhibition, caused by prior experience with a flavor not followed by postingestive effects, prevented conditioning when the flavor was later paired with gastrointestinal filling. Without the prior experience of sham feeding, the distinctively flavored solution conditioning did occur on the following real feeding tests. Both of these experiments thus provide strong support for the conditioning hypothesis originally proposed by Davis and Campbell (1973).

The extinction interpretation of the progressive increase in intake with sham feeding experience was tested in still another way by Davis and Smith (1990). We reasoned that the progressive increase in sham intake would not occur if real feeding intake tests were interspersed between sham feeding tests because the real feeding test would reestablish whatever extinction had occurred in the preceding sham feeding test. using 0.8M sucrose we tested this hypothesis by giving rats five cycles of a sham feeding test followed by two real feeding tests. With this testing procedure the rats showed greater intake on the sham feeding tests than on the real feeding tests but did not show a progressive increase in sham intake (Figure 3). However, as can be seen in that figure, when given five consecutive sham feeding tests after the five cycles of two real followed by a sham test, intake increased significantly.

An examination of the rate of licking during the real and sham feeding tests provided an explanation for this effect. On both types of test the rate of licking declined rapidly during the first four to five minutes, and on the real feeding tests it continued

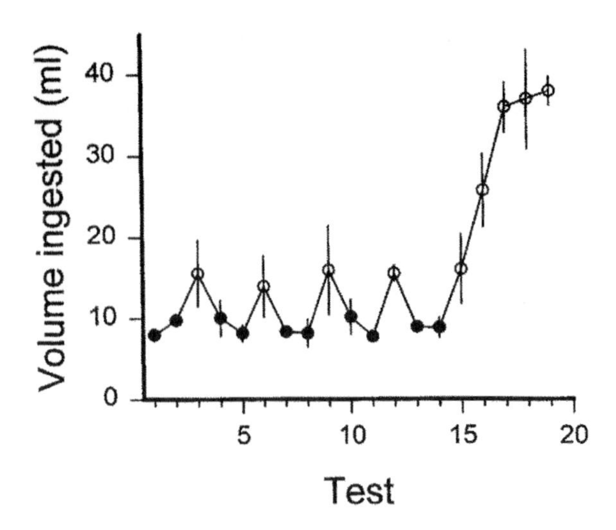

FIGURE 3 Volume ingested of 0.8M sucrose during 30-min real feeding (filled circles) and sham feeding (open circles) tests. Redrawn from Davis and Smith (1990). By permission of the American Physiological Society.

to decline rapidly to zero. However, on the sham feeding tests the licking rate remained substantially above zero until the end of the test. Figure 4 shows the rate of licking 0.8M sucrose on a real feeding test and on a sham feeding test following it. It is only from about 5 min on that the rates of licking are reliably different, and it is this difference that accounts for the greater intake on the sham feeding tests than on the real feeding tests. In our original study, when the rats were later given five consecutive sham feeding tests, intake increased progressively over the five tests because of an increase in the rate of ingestion early in the test, indicating that, without the intervening real feeding tests, sham intake would increase.

We interpreted the results of that study as supporting the view that the increase in intake that occurs on consecutive sham feeding tests does so because of the extinction of conditioned inhibitory control over the rate of licking during the first few minutes of the test. Furthermore, we confirmed the finding, first reported by Davis and Campbell (1973), that the principal change in the ingestive behavior of the rat that accounts for the rise in sham intake with experience is an increase in the rate of ingestion during the first part of the meal. The learned inhibitory control of ingestion that extinguishes with sham feeding experience occurs at the beginning of the meal. This type of conditioned satiety is to be contrasted with that reported by Booth (1972) and by Booth and me (1973), which exerts its effects at the end of a meal.

These studies by Weingarten and Kulikovsky and by ourselves pointed clearly to extinction of an acquired control over ingestive behavior as the explanation for the progressive rise in sham feeding that occurs with sham feeding experience. Because extinction of this control occurred when orophyrangeal stimulation was not followed by the usual gastrointestinal filling that occurs with real feeding, it was clear

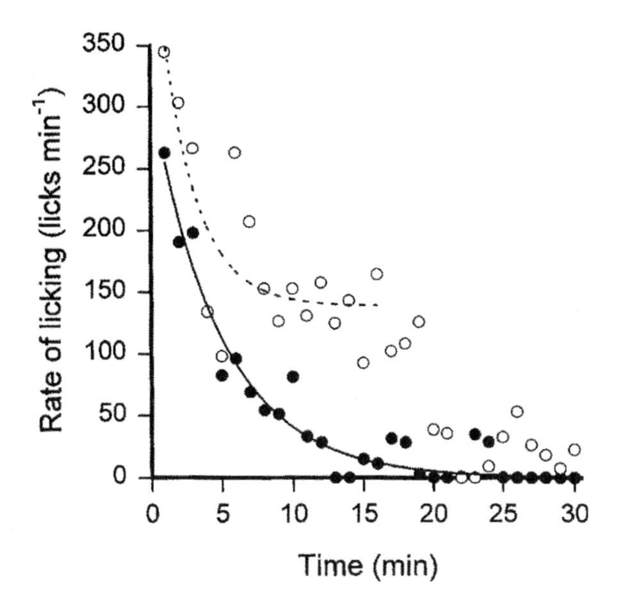

FIGURE 4 Rate of licking at 1-min intervals during a real feeding test and on the following sham feeding test. The data are plotted at 1-min resolution rather than at the 2-min resolution shown in the original. Redrawn from Davis and Smith (1990) by permission of the American Physiological Association.

that the unconditioned stimulus on which the conditioning depended had an origin in the gastric or postgastric (or both) compartments. Our next study was designed to decide among these alternatives. To do this we tested rats using 0.8M sucrose on the schedule of one sham test followed by two real feeding tests used in our previous study, but this time with the pylorus of the rats closed by the inflation of a pyloric cuff (Davis *et al.,* 1993). This confined all of the ingested fluid to the stomach during the real feeding tests. Furthermore, because the stomach was drained of all accumulated fluid at the end of real feeding tests before the pyloric cuff was deflated, there was no possibility for an oral-postgastric association to develop.

In this experiment the results were different. Intake increased progressively with experience on the real feeding tests as well as on the sham feeding ones (Figure 5). This indicated that postgastric stimulation plays an important role in the development of the conditioned control of ingestion because eliminating it on all tests led to a progressive increase in intake on the real feeding tests. Nevertheless, there was evidence of a role for gastric stimulation in the development of conditioned inhibition of ingestion as well because, although intake increased with sham feeding experience, it increased more slowly than when no real feeding tests intervened. Some stimulation associated with gastric filling on the intervening real feeding tests must have provided an unconditioned stimulus for the maintenance of an inhibitory control that prevented sham intake from increasing as rapidly as it would have otherwise.

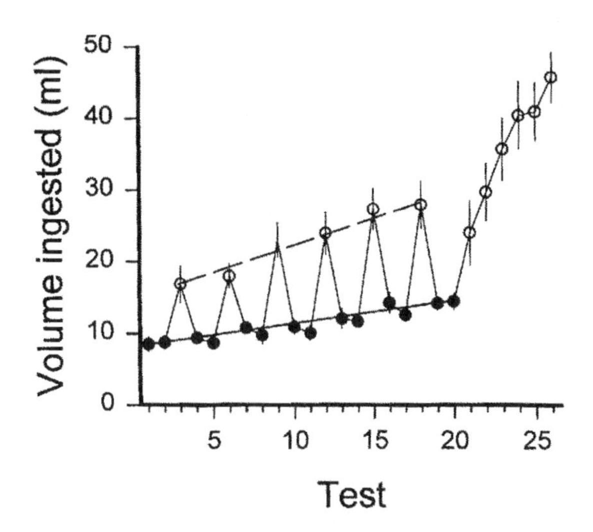

FIGURE 5 Intake of 0.8M sucrose on real feeding tests (closed circles) and sham feeding tests (open circles). Lines were added to the original to emphasize the trends in the data over trials. Redrawn from Davis *et al.*, (1993) by permission of the American Physiological Association.

Further evidence that stimulation arising from the stomach plays an important role in the acquired control of ingestion comes from a recently reported study using a different approach. If postgastric stimulation were the sole source of unconditioned stimulation, then the conditioned control of ingestion should extinguish if rats are given a series of real feeding tests but with the pylorus closed during the test. Rauhoufer and her colleagues (1991) tested this by giving rats a series of real feeding tests with 0.8M sucrose as the test solution. On all of the tests the pylorus was closed by inflating a cuff surgically implanted around the pylorus. At the end of each test all of the fluid that had accumulated in the stomach was drained, and the stomach was flushed with saline before the cuff was released. The results were that intake remained virtually constant during 14 consecutive cuff-closed tests. There was no indication of even a tendency for intake to increase or decrease during the 14 tests. Clearly some consequence of gastric filling in the absence of postgastric stimulation is sufficient to maintain conditioned inhibitory control of meal size. Only when both are eliminated, as in sham feeding tests, does this conditioned control of ingestion extinguish.

If meal size is controlled in part by a mechanism that extinguishes during consecutive sham feeding tests, then, when a rat is returned to a real feeding procedure following extensive sham feeding testing, meal size should be greater, at least on the first and, perhaps, the second real feeding tests. This follows from the fact that because the rate of ingestion at the beginning of a sham test does not decrease very rapidly it should remain high on the real feeding test that immediately follows a series of sham tests. Some time later the rate of ingestion will begin to decline, but not until the effects of gastrointestinal filling exert their usual negative feedback effects. If the reconditioning of the decline in the rate of ingestion is rapid, then meal size on subsequent real tests will return to normal, but the theory predicts a larger than normal meal on the first real feeding test following a series of sham tests and on any subsequent ones until conditioned control is reestablished.

The first study to confirm this was one designed to determine if rats develop a preference for real or sham feeding (Van Vort and Smith, 1983). In their study Van Vort and Smith gave their rats 18 cycles of alternating real and sham feeding tests with vitamin fortified sweetened milk. Each type of test was associated with a distinctively flavored milk and was conducted at a different cage location. At the end of this training they were given real feeding tests but with the flavor and location previously associated with sham feeding. On the first of these reversal tests the rats ingested significantly more than on the last real feeding test before the sham feeding experience. This hyperphagia, although statisically significant, was short lived for it had extinguished by the third trial of reversal training.

They (Van Vort and Smith, 1987) later replicated this finding in two follow-up studies. In one of them, in which the volume sham fed was limited to the volume real fed on the preceding test, they found that the increased intake on a real feeding test following a sham feeding experience was accompanied by an increase in intake during the initial 5 min of the test. In the second of these studies, in which the sham feeding tests were extended to 60 min, they found enhanced real feeding on the 2 real feeding tests immediately following the sham feeding experience. However, in this experiment they did not find an accompanying increase in the intake during the first 5 min.

In summary, evidence from a variety of different experiments and from different laboratories now makes it very clear that the progressive rise in intake that occurs when rats are given a series of consecutive sham feeding tests with concentrated test solutions occurs because of the extinction of acquired inhibitory control. This extinction can be prevented by interspersing real feeding tests between sham tests. The evidence is also clear that this inhibition exerts its influence on ingestive behavior at the beginning, rather than the end (Booth and Davis, 1973), of the test because it is at the beginning of a test that the changes responsible for the increase in sham intake with experience occur.

The possibility of conditioned control of ingestion was considered when we developed our original model, and an ingestion rate function incorporating it was derived (Davis and Levine, 1977). This rate of licking function had a new parameter, m, whose magnitude was assumed to depend on a conditioned association between the flavor of the test solution and some postingestive event. The rate of ingestion function, with this assumption incorporated became $l(t) = gpde^{(-d(rk+m)t)}$. When rats are given a sham feeding test for the first time, following real feeding experience with the same test solution, the value of m will be greater than 0 because of past orophyrangeal-postingestional conditioning and the rate of ingestion function will be $l(t) = gpde^{(-dmt)}$. However, as the rat gains sham feeding experience the value of m will decline toward 0 because of extinction, and the rate of ingestion function will be $l(t) = gpd$ when extinction is complete and $m = 0$. We reported evidence from a previous study (Davis and Campbell, 1973) that the value of the rate of decay parameter of the exponential function did decrease with sham feeding experience providing evidence to support that modification of the model.

The evidence supporting this idea is now much stronger, coming from different experiments performed in different laboratories, so this modification needs to be taken seriously, at least for the types of test solutions that have been used by the investigators studying this problem. It should be noted, however, that all of the test solutions that have been used so far to test this hypothesis have been relatively concentrated sucrose solutions, milk or a chemically defined diet. It is not yet known whether the phenomenon occurs with less concentrated sucrose solutions, other types of carbohydrates, fats, or with diluted mixed composition test solutions. Thus, further modifications of the model beyond the inclusion of a parameter to account for conditioned inhibitory control of ingestion at the beginning of tests using concentrated test solutions must await the results of future research.

2. Sham Feeding—Steady State

We now have a better understanding of why intake increases progressively as rats gain experience sham feeding concentrated solutions. This understanding provides confidence for introducing the parameter, m, in the model to account for the acquired control of ingestion rate at the beginning of a test. However, less is known about the reason for the other violation of the model's prediction concerning the rate of ingestion during sham feeding. This is the common finding that even in rats with extensive sham feeding experience the rate of ingestion declines during sham feeding tests. The model predicts that on any sham feeding test the rate of ingestion will depend only on an individual difference factor, d, the evaluation of the test solution, p, and the concentration of the test solution, g. That is, the rate of ingestion should remain constant for any rat on any sham feeding test regardless of the prior sham feeding experience of the rat. Evidence from a variety of laboratories indicates that this is not the case.

This effect was first seen in the results of the first study to use the open gastric fistula to study sham feeding (Young *et al.*, 1974). It can be seen in Figure 1 of their study that intake declined from about 14 ml per 5 min to about 8 ml per 5 min during the 2-hr test, even after the experience of two 2-hr sham feeding tests. In that study each sham feeding test was separated from the next by a real feeding test so it is possible that the decline occurred because the intervening real feeding tests prevented the complete extinction of conditioned control. Nevertheless, a similar effect has been reported in other studies in which rats had a number of consecutive sham feeding tests. For example, the effect is seen clearly in a study designed to investigate the effect of variation in sucrose concentration on sham intake (Weingarten and Watson, 1982). A decline in the rate of ingestion measured at 5-min intervals is most apparent with the most concentrated solution (40%) used but is clearly apparent with the others as well. Because no real feeding tests intervened between the sham tests in that study, the decline in the rate of ingestion during a sham feeding test must be due to some other influence on the rate of ingestion.

A study by Greenberg, Smith, and Gibbs (1990) done for reasons other than to demonstrate this phenomenon shows the decline in the rate of ingestion during a sham feeding test by experienced sham feeders clearly. In their study rats were trained to sham ingest BioServ, a chemically defined diet, until intake was stable (no more than 10% variation in intake for 3 days) before the experiment was begun. In Figure 6 it can be seen that intake measured at 5-min intervals under the saline control treatment of these highly experienced sham feeding rats declined in a steady and consistent manner over the 90-min test. The investigators were interested in other matters so they did not determine whether the decline was statistically reliable. However, the data are still available and, thanks to Dr. Greenberg's willingness to share them, were analyzed statistically. That analysis revealed that the decline in the rate of ingestion measured at 5 min intervals was statistically significant, $F_{17,102} = 7.7, p < 0.001$ and that there was a significant linear component in the trend, $F_{1,6} = 54.7, p < 0.001$. Ingestion declined at a slow but constant

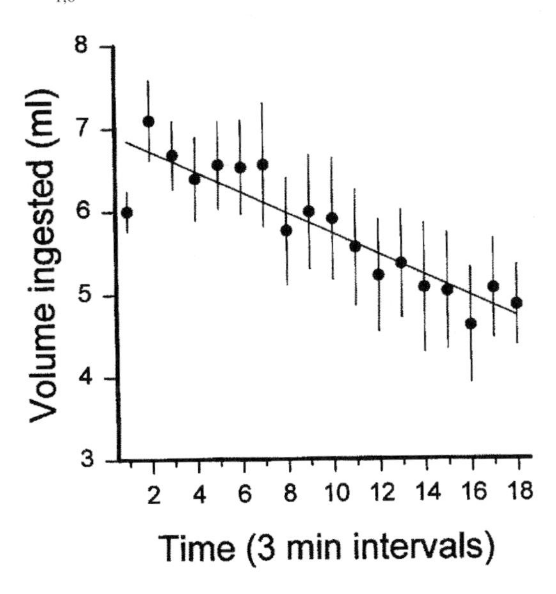

FIGURE 6 Volume of BioServ ingested measured at successive 3-min intervals during sham feeding. The scale below 3 ml has been truncated to emphasize the slope. Redrawn from Greenberg, Smith, and Gibbs (1990) by permission of the American Physiological Association.

rate throughout the 90-min sham feeding test, and, because these were highly practiced sham feeding rats, it is unlikely the decline can be attributed to acquired control of ingestion.

The reason for this decline in the rate of licking during sham feeding in rats with extensive sham feeding experience is not well understood primarily because it has not been studied extensively. There are a number of possible explanations. One is that postingestive stimulation is still playing a role mediated by esophageal, gastric, and possibly postgastric (Sclafani and Nissenbaum, 1985) stimulation by the ingested fluid. It seems unlikely that this is the explanation, however, because such stimulation must be very weak relative to its magnitude on real feeding tests where postingestive stimulation does play a role in controlling ingestion. Nevertheless, the possibility cannot be ruled out at this time.

Related to this is the possibility that some unconditioned effect of sham feeding is responsible. It was, in fact, to collect gastric secretions uncontaminated by ingested food that Pavlov developed the sham feeding technique originally. This cephalic phase response activates a variety of processes involved in digestion prior to the arrival of food in the gastrointestinal tract (Powley, 1977). Some of these responses to orophyrangeal stimulation such as the release of insulin (Louis-Sylvestre, 1978), glucagon (Geary, 1996), gastrin (Thein and Schofield, 1959), and HCl (Janowitz *et al.,* 1950), or others could act as negative feedback signals on the ingestion controller to reduce the rate of ingestion during sham feeding tests. This must be considered a highly speculative possibility, however, because there is no evidence I am aware of that they can serve this role.

Still another possibility is that the decline in the rate of ingestion is a consequence of the loss of the effectiveness of the test solution to maintain ingestion because of peripheral gustatory adaptation. In developing the model we assumed that this did not occur to any appreciable extent during a 30-min test, that is, we assumed that *gp* would remain constant during a test. This assumption was made to simplify the model, but the assumption may be wrong. Gustatory sensory adaptation does occur (Bartoshuk, 1968), and it may be sufficient to reduce the magnitude of the stimulation sustaining ingestion during a sham feeding test. After all, the gustatory apparatus is stimulated constantly by the test solution during a sham feeding test, and the longer the test is prolonged the longer the stimulation lasts. Nevertheless, I do not know if the magnitude of gustatory adaptation is sufficient to account for the magnitude of the decline in the rate of ingestion in experienced sham feeding rats, and, as with the other possibility, there is no evidence at this time that sensory adaptation occurs during this type of test.

Swithers-Mulvey and Hall (1992) have provided an argument and data supporting still another possible explanation for the decline in the rate of ingestion during sham feeding tests. They have argued that this decline is due to habituation and that habituation occurs during both real and sham feeding tests. In developing their argument they rejected the idea that peripheral gustatory adaptation is responsible for the decline in the rate of ingestion because they reported that a rat's decreased responsiveness to a test solution can last for up to 3 hr, which is a longer period of time than is necessary for gustatory sensitivity to recover.

In summary, although it is clear that a decline in the rate of licking in experienced sham feeding is a reliable phenomenon, the mechanism underlying it is a mystery. A number of possible explanations exist, but the phenomenon, although noted, has received no experimental attention so these possibilities remain purely speculative. Because the effect is small it probably does not contribute much to the decay in the rate of licking that occurs during real feeding. Nevertheless, it appears to be real and thus will have to be incorporated into a more fully developed model of the control of meal size.

3. Real Feeding

In the original model negative feedback was represented by the combined parameters, *drk.* These letters referred to an individual difference parameter, *d,* and a neural representation, *k,* of gastrointestinal filling, *r.* We chose the letter *r* because we assumed that negative feedback was proportional to the volume ingested that was retained in the small intestine during the test. Our idea to base negative feedback on gastrointestinal filling came from our finding (Davis *et al.,* 1975) that adding mannitol, a sugar alcohol that is not absorbed from the intestine, to a test solution of Na saccharin and glucose reduced intake in proportion to its concentration. Our analysis of those data revealed that the reduction in intake could be attributed entirely to postingestive effects because mannitol was shown to have no effect at any concentration on the initial rate of licking, but increasing the concentration of mannitol in the test solution increased the value of *drk* linearly with the log of its concentration.

We assumed that, because it was not absorbed, mannitol would, by creating an osmotic gradient, create an accumulation of fluid in the intestine. Barbara Collins and I (Collins and Davis, 1978) confirmed this later by measuring the volume of the stomach and the small and large intestines in rats that had ingested a solution containing mannitol. We found that there was an accumulation of fluid in the small and large intestines, but not in the stomach following the ingestion of a solution containing mannitol with Na saccharin and glucose. This mannitol-induced accumulation of fluid in the intestines was later confirmed by Bernstein and her colleagues (Bernstein and Vitiello, 1978).

Carmen Perez, Dora Kung, and I (Davis *et al.,* 1994a) late showed in a different way that fluid accumulation in the small intestine can act as a negative feedback signal. We showed that when Acarbose is added to a 0.3M sucrose test solution intake is reduced not because of any effect of it on the flavor of the solution but because of an increase in the magnitude of negative feedback. Acarbose is an α-glucosidase that prevents the hydrolysis of disaccharides, and we showed, by measuring gastrointestinal content, that adding it to the sucrose solution caused an accumulation of fluid in the small and large intestines but not in the stomach.

Thus, although there is evidence supporting the idea that intestinal filling per se can serve as a negative feedback signal, it is now clear that was a very limited view. For example, there is evidence that the small intestine is the source of a different kind of negative feedback that controls water intake by water deprived rats. Hall (1973) first showed this in a study describing the construction and use of a pyloric noose to confine ingested fluid to the stomach during an ingestion test. In that study he reported that water deprived rats ingested more water in a 30-min test when the noose was closed, preventing postgastric stimulation, than when it was open. Because the rats drank more water in the absence of postgastric stimulation than with it, some signal, activated when the ingested water was being absorbed in the proximal intestine, must have acted as a negative feedback signal to limit water intake.

We recently confirmed that finding (Davis and Sayler, 1997) and, in addition, showed that confining isotonic saline to the stomach during the test reduced, rather than increased, intake. When water deprived rats ingest isotonic saline, intestinal stimulation may ultimately terminate ingestion, but, if so, it must be by a different mechanism than that operating when pure water is being ingested. Furthermore, if postgastric stimulation does contribute a negative feedback signal it seems unlikely to depend on intestinal distention because isotonic saline must be absorbed very rapidly.

There is now good evidence that the components of complex diets are likely to serve as negative feedback signals as well. Connie Campbell and I showed some time ago that duodendal infusion of glucose could suppress licking rate (Campbell and Davis, 1974). More recently Greenberg and her associates (Greenberg *et al.*, 1988; Greenberg *et al.*, 1990; Greenberg *et al.*, 1986) have demonstrated that intraduodenally infused fat in various forms can suppress intake in sham feeding rats, indicating a likely role as negative feedback for fats of fatty acids. Walls and colleagues (1995) have recently reported that a variety of nutrients such as glucose, maltose, L-phenylalynine, lsocal, and oleic acid when unfused into the duodenum at an appropriate concentration suppressed intake. Reidelberger (Reidelberger *et al.*, 1983) has also shown that duodenal infusions are capable of reducing intake.

All of these studies make clear that the origin and type of stimulation giving rise to negative feedback, while not fully understood, will most likely turn out to be multifaceted with gastric factors (Deutsch, 1990), intestinal factors, and peptides (Gibbs and Smith, 1992) all playing a role. This could be the reason my colleagues and I have shown that the value of *drk,* estimated by curve fitting, increases with the concentration of sucrose (Davis and Perez, 1993b), of corn oil (Davis *et al.*, 1995), of Polycose (Davis, 1996a), and, in unpublished studies, of Maltooligosaccharide and milk. The estimates of *drk* for the milk dilution study are listed in Table 1. These are all cases where increasing the concentration of the test solution would be expected to increase the magnitude of negative feedback because of a correlation between the concentration of the solution on the one hand and the colligative properties, caloric value, and chemical composition of the solution on the other hand, all of which are good candidates for negative feedback signals. When variation of the latter is minmized by varying the concentration of Na saccharin in a Na saccahrin + glucose mixture there is virtually no variation in the value of drk even when intake varies over a wide range (Breslin *et al.,* 1996).

In summary, it is now clear that placing the entire burden of negative feedback on a signal derived from intestinal filling alone was much too limited and short sighted. Strong claims for the stomach as the primary origin of negative feedback have been made (Deutsch, 1990), some of which are being questioned (Phillips and Powley, 1996), and there is ever expanding support for the view that peptides such as CCK (Smith and Gibbs, 1992), bombesin, and others (Gibbs and Smith, 1992), and specific nutrients serve as negative feedback signals. Nevertheless, the idea that negative feedback, of one kind or another, acts to control ingestion by controlling the rate of decline of the rate of licking appears to be supported by results from a variety of studies all showing that the magnitude of this parameter increases with the concentration test solutions. This is true whether the test solution contains a single compound or is a mixture of a variety of nutrients. Thus, if *drk* is seen as a generalized feedback signal, not necessarily dependent on intestinal filling alone, the evidence supports the role given it by the model as a control over the rate of licking during a test.

B. Initial Rate of Ingestion

1. Introduction

Because the value of *gpd* in the rate of licking function, $l(t) = gpde^{-drkt}$, is the rate of licking at time $t = 0$ ($gpde^{-drk0} = gpd$), the model predicted that the initial rate

TABLE 1
PARAMETERS OF THE WEIBULL FUNCTION

Milk Dilution	A	B	C	$t_{1/2}$
1:1	226.0	0.089	1.5	8.8
SEM	16.3	0.008	0.26	
2:1	208.9	0.083	1.7	9.7
SEM	12.9	0.005	0.26	
4:1	180.6	0.066	3.0	13.35
SEM	11.4	0.004	0.68	
8:1	177.6	0.062	3.9	14.7
SEM	13.3	0.004	1.3	
16:1	140.4	0.043	2.9	20.5
SEM	10.2	0.003	0.36	

Note. Estimates of the parameters and their standard error of estimates (SEM) for the Weibull function, $ae^{-(bt)^c}$, fit to the rate of licking functions for five dilutions (water:milk) of sweetened condensed milk.

of licking a test solution would be a joint function of the concentration of the test solution, *g*, its evaluation by the subject, *p*, and an individual difference parameter, *d*. Because the magnitude of *gpd* reflects the values of *g*, *p*, and *d* it is necessary to keep two of them constant while varying the other in investigating the contribution of each to the magnitude of the whole. It is always safe to assume significant individual differences in any measurement of animal behavior so experiments conducted to measure *g* or *p* must be within subjects designs. In this way *d* is kept constant, and variations in the dependent variable can be attributed to variations in *g* or *p* depending on which is being manipulated.

We interpreted the numerical value of the initial rate of licking to be a measure of the magnitude of the input to the system in the CNS of a particular rat that drives its ingestive behavior. We also interpreted *gpd* a being a measure of the palatability of a test solution for that animal because it referred to what we presumed to be the determinants of that experience or sensation. I do not feel now that that second interpretation was wise because it attributed something to the rat that can never be measured and, therefore, never be tested quantitatively. It also deflected attention from what I consider to be the principal aim of the model, which is to provide a method for quantifying the ingestive behavior of the rat in such a way that quantitative relationships can be established between its behavior and the neural activity in the brain responsible for it.

The concepts of palatability and hedonics were applied to the rat, most notably by P. T. Young and his colleagues (Young, 1966), and continue to be today, as an explanation for the variations in intake that accompany variations in the concentration and composition of diets. It must be kept in mind though that these are subjective concepts that can only be quantified by psychophysical scaling methods applied to human verbal responses or representations of them. They can be used, therefore, in dealing with nonverbal species only as metaphors, not as variables that can be quantitatively related to variation in some experimental manipulation. For a quantitative analysis of the ingestive behavior of the rat, a subjective term such as *palatability* is useless and can be dispensed with entirely without jeopardizing the analysis.

However, the term *palatability* is in common use in our field, and it does refer, in a loose way, to something that we believe has an influence on how much we eat. Seen in this way, as a nonquantitative term that we, as humans, can identify subjectively and apply metaphorically to nonverbal animals, it can be used as a short-hand way to describe the reaction to the flavor of a test solution or diet that, in part, determines how much of it will be eaten. It can be useful, that is, as long as it is used metaphorically and is not seen as something that can be quantified when dealing with nonverbal subjects. Because the purpose of our model was to quantify the behavior of a nonverbal animal, I will not attempt to evaluate the initial rate of licking as a measure of palatability. Rather I will focus on reviewing the evidence relevant to the idea that *gpd* is a measure of the evaluation of the magnitude of orophyrangeal stimulation by the subject and that it serves as an input that drives the ingestive behavior of the rat.

2. Measuring the Initial Rate of Ingestion

a. Curve Fitting Method. To evaluate *gpd* as a measure of an excitatory input to the CNS driving ingestive behavior, it is necessary to obtain a numerical value for *gpd.* This has been done in three ways, each of which has advantages and limitations. One approach obtains estimates of *gpd* by curve fitting. The function $y = ae^{-bt}$ can be fit statistically to a rate of ingestion or a rate of licking function, and the resulting estimate of the initial rate parameter, a, can be used as a measure of *gpd.*

An estimate of the parameter a can also be obtained by fitting the cumulative form of the rate of licking function, $C(t) = ab^{-1}$ $(1\text{-}e^{-bt})$, to the cumulative intake function. This provides an estimate of the asymptote $ab-^{1}$ and of b from which an estimate of a can be obtained by division.

One problem that can arise in obtaining an estimate of *gpd* from curve fitting is that, because the estimate of both of the parameters of the equation is determined by all the data in the curve, deviations in the data from a true exponential rate or cumulative curve can distort the estimates. If, for example, the rate of licking during the initial peirod of the test falls very rapidly the estimate of y at $t = 0$, the estimate of the initial rate of ingestion, will be greater than if the rate of licking falls more slowly because the data in the entire curve is used in the curve fitting algorithm. Therefore, in using curve fitting to obtain these estimates the goodness of fit has to be examined for significant deviations from an exponential function.

One way of doing this is to use the Weibull function $y = ae^{-(bt)^c}$ to fit the function. The Weibull provides the shape parameter c which is a measure of the extent to which the function deviates from a true exponential. When $c < 1$, the function decays more rapidly than would a true exponential. When $c > 1$, it has a shoulder, its initial decline is less steep than a true exponential. When $c = 1$, the function is the same as a simple exponential. Therefore, if this function is used to fit the rate of licking function, deviations from the true exponential can be quanitifed at the expense of estimating only one additional parameter. The value of c can then be used to determine whether the estimate of the initial rate is likely to be a good one or not.

Another problem, which has only become apparent recently, is that the magnitude of the estimate of the initial rate parameter will depend on the size of the time intervals used in constructing the rate of licking or ingestion curves. In our original, and in my more recent work, the number of licks occurring in successive 1 min intervals has been used to construct the rate of licking curves to which the curve fitting algorithms have been applied. The selection of 1 min as the integration interval was chosen because it was felt that this provided a fine enough resolution to capture the essential features of the curve and at the same time provide sufficient data for accurate curve fitting. The 30 data points obtained in a 30-min test session seemed to be more than enough for determining accurately the values of the parameters that determine the shape of the curve.

This causes no problems if one is interested in the average rate of ingestion at various times during the test. The parameter *gpd,* however, is not an average, it is the rate at which the process begins, the limit of the rate of ingestion as the time interval just after the test starts becomes smaller and approaches 0. If there is no change in the rate of ingestion during the first minute of the test, then the average of that interval will be the same as the instantaneous rate. If, however, there is a change in the rate during this interval, say a decline, then the average rate over the 1-min interval will underestimate the initial rate of ingestion, and the average will be less than the actual initial rate the longer the integration interval is. It is now known that the rat's rate of licking does decline significantly during the very early moments of testing (Smith *et al.,* 1992) so the duration of the interval over which the rate of licking is integrated will make a difference in the estimate of the initial rate. The longer the interval, the lower and the farther from the true value the estimate of the actual initial rate will be.

Fortunately, two other methods for obtaining a measure of the magnitude of the stimulus that drives ingestion are available and can be used to validate results obtained by the curve fitting method or as an alternative to it. Both of these are based on the assumption that if the influence of postingestive effects on ingestion are eliminated, or minimized, then the rate of ingestion measured over a short time interval will provide an estimate of *gpd.*

b. Brief Contact Method. The brief contact method, pioneered by P. T. Young (1967), offers the rat a test solution for a brief period of time, 30 sec to 2 or 3 min, and the volume ingested or the number of licks that occurred during that interval is recorded. The rationale for this approach is that if access to the test solution is restricted to a brief enough period of time, then the test will have been completed before postingestional consequences of ingestion will have had a chance to influence the measure. This brief contact method, therefore, should reflect only the stimulating effectiveness of the test solution and its evaluation by the rat. The advantage of this approach is that testing requires little time, and many different concentrations of a test solution can be evaluated in a single test session (Smith *et al.,* 1992). The validity of this approach depends, of course, on the assumption that during this brief period of time the rate of licking or ingestion depends only on the concentration of the test solution, and its evaluation and is uninfluenced by postingestional stimulation, sensory adaptation, or habituation. If the test period is brief enough, 60 sec or less, this assumption is a reasonable one. It is important to keep in mind though that this is only an assumption because I am not aware of any research that has attempted to determine how soon after a rat begins to eat the postingestional consequences of this activity begin to influence ingestion.

c. Sham Feeding Method. Another approach, based on the same assumptions as the brief contact method, is to use the sham feeding technique and measure the volume ingested or number of licks during a fixed period of time. If rats have been adapted to the sham feeding procedure for a sufficient number of tests so that any acquired controls over ingestion have extinguished (Davis *et al.,* 1993; Davis and Smith, 1990), then the volume ingested during the test will reflect only the stimulating effectiveness, and its evaluation of the test solution will be unaffected by postingestive or learned influences on ingestion. This method is somewhat more laborious and time consuming than the brief contact method, but if the rat is well adapted to the sham feeding procedure, it can be assumed that learned influences dependent on orophyrangeal-postingestive associations will have been extinguished. Furthermore, if a pyloric cuff is used in conjunction with sham feeding then postgastric influences on ingestion (Sclafani and Nissenbaum, 1985) can be ruled out with certainty.

There are some disadvantages with both methods. With the brief contact method the assumption that learned influences and postingestional ones have no influence on the behavioral measure may not be valid. The role, if any, of acquired control of ingestion in the brief contact method has not been investigated so it is unclear at this time how free of this influence the measure may be . The sham feeding method requires surgery, and testing is time consuming. Furthermore, by extending the test to 30 min or more, effects on ingestion such as habituation (Swithers-Mulvey and Hall, 1992) and gustatory adaptation (Bartoshuk, 1987) have an opportunity to affect the measurement.

3. Variation of Stimulus Type and Intensity

Many more studies have measured variations in *gpd* in response to variations in *g,* the concentration of the test stimulus, than in response to variations in *p,* the evaluation of the input by the rat. Therefore, I will review first the studies that have measured *gpd* when the type and concentration of the test solution was varied and then discuss the few studies that examined variations in *p.*

The supporting data we (Davis and Levine, 1977) presented in 1977 used the curve fitting approach to data obtained under a variety of different test conditions to see if the estimates of the initial rate of licking varied in ways that were consistent with expectation. One approach was to obtain estimates of *gpd* under conditions where intake varied widely, but where the test solution remained constant so we could be reasonably certain that its evaluation by the animal did not change. In one of those tests the initial rate of ingestion was estimated from the cumulative intake curves obtained from rats ingesting milk on a real feeding test and on five consecutive sham feeding tests. On those tests the volume ingested increased progressively from 12 ml on the real feeding test to 91 ml on the last sham test, but the estimate of the initial rate of ingestion was essentially constant, varying nonsystematically from a low value of 1.7 ml min^{-1} (first sham test) to a high value of 2.7 ml min^{-1} (real test). This relative constancy of the estimate of *gpd* was expected because the test solution was the same on all of the tests, and we did not expect sham feeding experience to affect its evaluation by the rats. The fact that it did remain constant confirmed our expectations.

Another test examined the estimates of *gpd* obtained in a series of six tests where rats were offered a saccharin + glucose mixture to which increasing amounts of mannitol were added. We had shown in an earlier paper (Davis *et al.,* 1975) that adding mannitol in this concentration range to a saccharin + glucose solution had no systematic effect on the number of licks during a 3-min brief contact test. In that study the number of licks in the 3-min tests ranged unsystematically from a low of 456 (152 licks min^{-1}) at 0.075M to a high of 532 (177 licks min^{-1}) at 0.4M suggesting that it was tasteless or neutral for the rats. When mannitol in concentrations from 0 to 0.5Min 0.1M increments was added to the saccharin + glucose solution and the rats were tested for 30 min, the total number of licks recorded during the test decreased from 3607 to 885 licks. Nevertheless, the estimates of *gpd,* obtained from fitting *C(t)* to the cumulative licking functions, varied in no systematic way from a low of 202 licks min^{-1} to a high of 229 licks min^{-1}.

The estimates of the rate of licking obtained from curve fitting were greater than those obtained in the brief contact method because the estimates obtained by the former method were estimates of the instantaneous initial rate of licking and by the latter were averages over a 3-min period during which the rate of licking declined. With this in mind, the estimates obtained by the two methods can be seen to be in good agreement, providing support for the view that the initial rate of ingestion, as estimated from the curve fitting method, and the brief contact method provide quantitatively similar results.

We also used the curve fitting method to measure variations in *gpd* where we had reason to believe that *gpd* would vary significantly. Using the brief contact method and a lickometer, Young and Trafton (1964) had shown that when quinine is added in increasing concentration to a series of different concentrations of sucrose the number of licks made by the rat during a 1-min test period decreased monotonically with increasing quinine concentration. This occurred with all of the sucrose concentrations, and the effect was more pronounced with the weaker concentrations of sucrose. To evaluate the effect of quinine on *gpd,* we varied the concentrations of quinine, ranging from 2mM to 20mM, added to a 0.3M glucose solution and measured the rate of licking during 30-min test sessions.

In our study the total number of licks decreased with increasing quinine concentrations from about 2,000 when 2mM quinine was added to about 500 when 20mM quinine was added to the glucose. The estimates of the initial rate of licking, obtained by fitting *C(t)* to the cumulative licking function decreased linearly from 205 licks min^{-1} at the lowest concentration of quinine to 61 licks min^{-1} at the highest. Because Young and Trafton (1964) used different concentrations of sucrose as the base solution and different concentrations of quinine, an exact comparison between their data and ours cannot be made. Nevertheless, at the lowest concentration of sucrose they used, which would be most comparable to the 0.3M glucose we used, the values of the rate of licking they reported fell from about 200 licks min^{-1} to about 25 licks min^{-1} over the concentration range they used. The values we reported ranged from about 205 licks min^{-1} to about 60 licks min^{-1} over the quinine concentration range we used. The agreement between these two studies is thus quite respectable.

The effect of quinine adulteration on a sucrose test solution with sham intake has also been reported by Weingarten and Watson (1982). They reported that the sham intake of a 30% sucrose solution was significantly decreased from about 27 ml to about 14 ml when 0.0025% quinine was added, and further decreased to about 3 ml when 0.005% quinine was added. They also measured intake at successive 5-min intervals during the tests and found that the reduction in intake was evident during the first 5 min, although this was much more apparent with the highest concentration of quinine.

We also tested the model with solutions that we assumed would increase, rather than decrease, *gpd* with increasing concentration. To do so we fit the *C(t)* function to cumulative licking functions obtained from rats ingesting five concentrations of maltose ranging from 0.1M to 1.6M. The total number of licks in the 30-min test session decreased with concentration from 5,000 to 2,088 over this concentration range but the estimates of *gpd* increased from 266 licks min^{-1} at 0.1M to 328 licks min^{-1} at 1.6M. We also reported that the number of licks during 3-min brief contact tests with maltose increased in a linear fashion from about 140 licks min^{-1} to about 200 licks min^{-1} with concentrations from 0.1M to 1.6M in 3-min tests. Although the estimates obtained from the 3-min tests were less than those obtained from curve fitting, the trends were the same. The estimates obtained from the 3-min tests are lower than those obtained from curve fitting because, as in the mannitol study, the former measure was obtained by averaging over a 3-min period during which time the rate of licking decreased. The similarities in the order of magnitudes of the two estimates and in the trends again provided encouragement that the two approaches generated similar results.

Joyner and her colleagues (Joyner *et al.,* 1985) reported that the sham intake of four common sugars increased monotonically with concentration. They reported that the 30-min sham intake of maltose in nonfood-deprived and 17-hr food-deprived rats increased monotonically over a concentration range of 0.1M to 0.8M (Figure 7). Note that, as in our report, with 0.1M maltose, the lowest concentration of maltose used in either study, the estimates of *gpd* were relatively high, 140 licks min^{-1} in our study and about 6 ml (nondeprived) and about 25 ml (food-deprived) in Joyner's study. Joyner and Smith (1986) have reported similar results in both lean and obese Zucker rats.

A recently published study Breslin, Rosenak, and me (Breslin *et al.,* 1996) investigated the effect on ingestive behavior of increasing the concentration of saccharin in a saccharin + glucose mixture. In that study rats were offered, in 30-min test sessions,

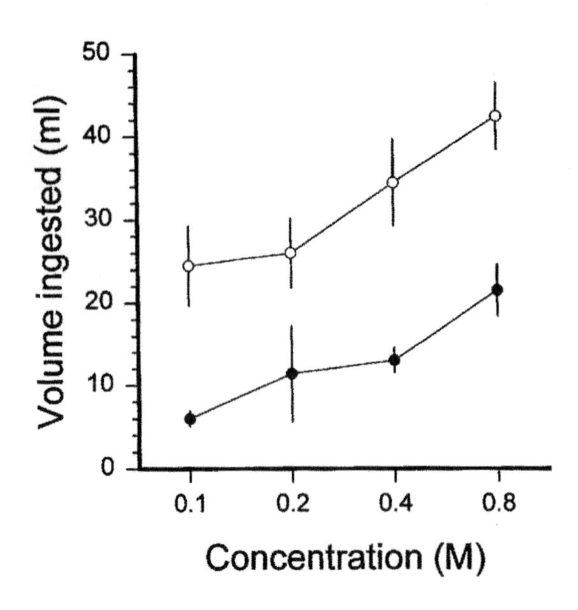

FIGURE 7 Volume ingested during 30-min sham feeding tests as a function of maltose concentration under 0-hr deprivation (filled circles) and 17-hr food deprivation (open circles). Redrawn from Joyner et al. (1985), by permission.

a 0.2M glucose solution to which 6 concentrations of Na saccharin ranging in from 0.25mM to 8mM were added. Total intakes ranged from about 7 ml (0.2M glucose alone) to about 27 ml (8mM Na saccharin + 0.2M glucose). Associated with this increasing intake with Na saccharin concentration was a closely correlated increase in the estimates of the initial rates of licking from about 70 licks min^{-1} with glucose alone to about 280 licks min^{-1} at 8mM Na saccharin + 0.2M glucose.

I am not aware of any studies using these combinations of saccharin and glucose with the brief contact or sham feeding techniques so a comparison is not possible. Young and Trafton (1964), however, reported a study using the brief contact method with mixtures of Na saccharin and sucrose and reported that adding 6mM Na saccharin to 1% sucrose increased the number of licks min^{-1} over that obtained with 1% sucrose alone. There was no further increase as the saccharin concentration was increased still further to 48mM, and there was a decline with saccharin concentrations of 100mM and 200mM. Their finding of an increase in brief contact lick rate with the addition of 6mM Na saccharin to 1% sucrose is similar to our (Breslin *et al.,* 1996) report, but an exact comparison is not possible because they (Young and Trafton, 1964) used much higher concentrations of saccharin and mixed it with sucrose rather than glucose. I am not aware of any brief contact or sham feeding tests with saccharin + glucose mixtures so it is not possible to compare our finding using the curve fitting method with the other two methods of measuring the magnitude of the input signal to the control system.

In the case of sucrose it is possible to make a comparison among all three methods used for studying the magnitude of the input signal to the ingestion control system. We (Davis and Perez, 1993b) reported that the estimates of the initial rate of licking obtained by the curve fitting method increased from 58 licks min^{-1} with 0.05M sucrose to 363 licks min^{-1} when the rats were tested with 0.4M sucrose. There was no further increase in this estimate when the rats were tested with 0.8M sucrose. A number of studies (Young and Trafton, 1964; Smith *et al.,* 1992; Davis, 1973) using the brief contact method have shown that the rate of licking increases rapidly with concentration at the low and intermediate concentrations of sucrose and then increases slowly or remains constant at concentrations around 0.4M and above.

Three sham feeding studies have reported the results of variations in sucrose concentration on intake (Weingarten and Watson, 1982; Joyner *et al.,* 1985; Joyner and Smith, 1986). All three are consistent in showing that the volume ingested during 30-min sham intake tests by food-deprived rats increases monotonically with concentrations from 6% to 40% (Weingarten and Watson, 1982) and from 0.1M to 0.8M (Joyner *et al.,* 1985; Joyner and Smith, 1986). In one of the studies (Weingarten and Watson, 1982), in which intake was recorded at 5-min intervals throughout the tests, intake during the first 5 min was linearly related to concentration.

One discrepancy between the curve fitting method and the brief contact method should be mentioned, however. In an early study using the brief contact method (Davis, 1973) I studied the effect of increasing the concentration of sucrose on the licking rate of rats under a variety of different durations of food deprivation from 0 hr (immediately after a meal) to 48 hr. Food deprivation had virtually no effect on the brief contact measure at any of the concentrations used. In our more recent study (Davis and Perez, 1993b) where we used the curve fitting method we reported that 24 hr of food deprivation increased the estimates of *gpd* with the 3 lowest concentrations of sucrose, but had no effect on the estimates of the 2 highest. Whether the lack of an interaction between deprivation and concentration with the brief contact method and the presence of one with the curve fitting method in the estimating *gpd* represents a serious challenge to the idea that the three approaches are measuring the same thing will only be resolved by future research. For the present, the more frequent occasions of agreement among the three measures are grounds for encouragement.

In all the studies discussed so far, the measures of *gpd* have been monotonically related to the concentration of the test solution. With NaCl the relationship is nonmonotonic, providing an exception to the rule that the value of *gpd* is monotonically related to the concentration of the test solution. Breslin and his colleagues (Breslin *et al.,* 1993) have reported that the number of licks during the first 3 min of a 30-min test by rats in sodium balance increased with increases in NaCl concentration of the test solution from 0.05M to 0.16M and then decreased as the concentration was increased still further to 0.5M. Figure 8 shows the effect. When the same rats were later tested in a state of sodium deficiency induced by furosemide treatment, the same nonmonotonic pattern with a peak at 0.16M was observed, but the curve was elevated by a factor of at least 3 (Figure 8).

Young and Trafton (1964) studied the effect of adding NaCl in concentrations of 1%, 2%, 4%, and 8% to 6 different concentrations of sucrose ranging from 1% to 32% with the brief contact method. They reported that adding 1% Nacl to 1% sucrose increased the number of licks during the 60-sec brief contact tests and that further increases in the NaCl concentration resulted in decreases in the measure. With the higher concentrations of sucrose increasing their NaCl content to 2% had no effect on the number of licks, but increasing it beyond that led to decreases in the number of licks.

In the cases discussed so far there is a known gustatory receptor for the stimulus whose intensity in the test solution was varied. These stimuli, therefore, provided a good test for the assumption of the model that gustatory receptors were the transducers that activate the CNS ingestion controller. There are, however, a number of substances that are very effective in stimulating ingestion and for which there is no known gustatory receptor. If variations in their concentration causes variations in the estimates of *gpd* similar to those previously reported, then the assumption that gustatory stimulation, per se, that is, *g* in the model, is a necessary input for driving the ingestion controller is wrong. The substances I know of now that provide this challenge to the model are corn oil, and the mixtures of oligosaccharides, Polycose and Maltooligosaccharide.

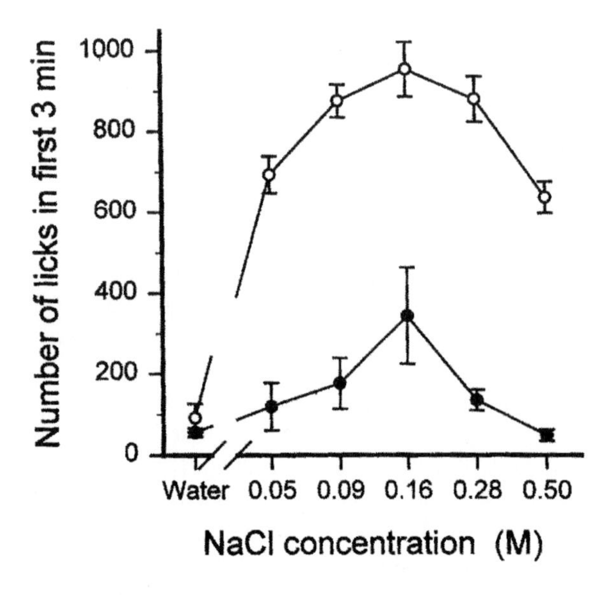

FIGURE 8 Number of licks during the first 3-min as a function of NaCl concentration under Na repletion (filled circles) and Na depletion (open circles). Redrawn from Breslin et al. (1993), by permission of the American Physiological Association.

In a recent study Theorora Kung, Rachel Rosenak, and I studied the effects of variation in the concentration of corn oil emulsions on the ingestive behavior of rats (Davis *et al.,* 1995). We measured the rats rate of ingesting these emulsions ranging in concentration from 0.125% to 64% in 30-min tests. Estimates of the initial rate of licking, obtained by fitting the $l(t)$ function to the licking curves, were a positive linear function of concentration over the entire range tested. Mindell and colleagues (Mindell *et al.,* 1990) reported that intake in 30-min sham feeding tests was an increasing linear function of concentration over the same range. Thus the sham feeding and initial rate of licking measures obtained from curve fitting are in essential agreement for corn oil. Although there is no known gustatory receptor for corn oil, increasing its concentration increases both sham intake and the initial rate of licking. This suggests that for the rat there is a nongustatory mechanism governing the activating effect corn oil has on the ingestion controller. The implication of this for the model is that variation in gustatory stimulation, g, is not the only thing that causes variation in the value of gpd. Some other parameter responsive to some unknown orosensory stimulus is needed to account for these findings.

The same seems to be true for another group of substances that can stimulate vigorous ingestive behavior, but have no known gustatory receptors. Polycose is a complex mix of polysaccharides, which although reported to be relatively tasteless (Feigin *et al.,* 1987) or slightly sweet (Hettinger *et al.,* 1996) to humans, is highly effective in stimulating ingestion in rats (Sclafani, 1987). Sclafani and his colleagues have reported that increasing the concentration of Polycose increases the number of licks made by rats during 3-min brief contact tests (Sclafani and Clyne, 1987) and that intake increases with concentration when Polycose is used in sham feeding tests (Nissenbaum and Sclafani, 1987). Both of these results indicate that increasing the concentration of Polycose in a test solution increases its effectiveness in stimulating ingestion in rats, and, therefore, should increase estimates of gpd obtained by the curve fitting method. Recently I have reported (Davis, 1996a) that the estimates of the initial rate of licking obtained from fitting an exponential function to the rate of licking curves obtained from 30-min access tests with Polycose ranging in concentration from 0.6% to 40% is a linear function of the log of the concentration. Here again the results obtained from the brief contact method and the sham feeding method are in substantial agreement with those obtained by estimating the initial rate of licking by the curve fitting method.

In an unpublished study I examined the effect of variation in the concentration of Maltooligosaccharide on the rat's rate of ingestion. Maltooligosaccharide, like Polycose is a mixture of oligosaccharides, but, unlike Polycose, it contains no glucose or maltose. In order to compare the effect obtained with Maltooligosaccharide with the missing mono- and disaccharides, I included an analysis of glucose and maltose as well. Figure 9 shows the initial rate of licking glucose, maltose, and Maltooligosaccharide in 30-min test sessions as a function of concentration. The

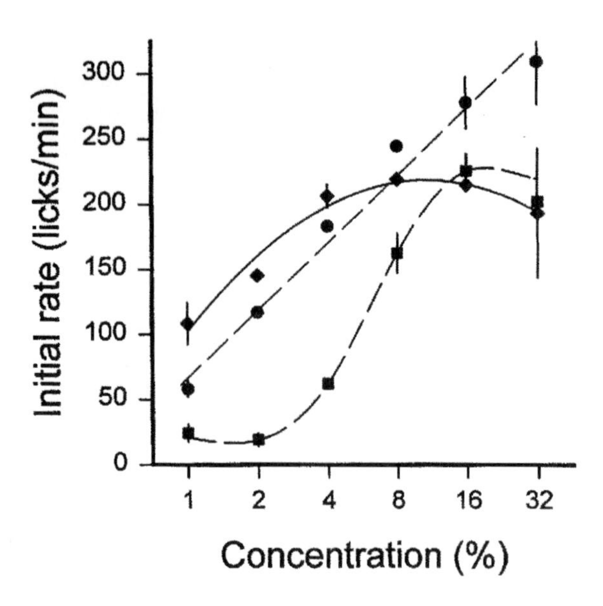

FIGURE 9 Initial rate of licking for glucose (squares), maltose (circles), and Maltooligosaccharide (diamonds).

data were obtained by the curve fitting method. Although the shapes of their dose-response functions are different, it is clear that in each case there is an increasing trend in the estimates of this parameter of the licking rate curve.

In summary, these studies are all consistent with the idea that variations in the concentration of substances in solutions that are acceptable to the rat result in variations in the rate at which the solution will be ingested when first tasted. In all but the case of NaCl and Na saccharin the relationship is linear with the log of the concentration of the substance. In the odd cases the relationship has the form of an inverted V-shaped function. With these two substances it appears that there is an optimal concentration above which increasing the concentration further introduces, perhaps in the case of Na saccharin, a bitter component, and in the case of NaCl an aversive element.

When the model was developed the letter *g* was chosen because it was assumed that stimulation of the gustatory receptors was responsible for stimulating intake. We now have clear evidence previously discussed that there are at least two exceptions to that rule. The initial rate of ingestion of corn oil and the two polysaccharides, Polycose and Maltooligosaccharide, is a monotonic increasing function of log concentration of the emulsion or solution. This pattern is identical to that obtained with the mono- and disaccharides, for which there are known receptors yet there are no known gustatory receptors for corn oil or the polysaccharides. Clearly the meaning of the symbol *g* in the expression *gpd* can no longer be taken to refer to the classical gustatory receptors alone. Some other orosensory system is responsible for providing input to the CNS that drives the ingestion of corn oil and the polysaccharides in a way that is similar to that provided by the classical "sweet" and Na gustatory receptors.

Perhaps the letter *o* (for orosensory) should be substituted for *g* (for gustatory) in the term *gpd.* At any rate, it is now clear that the ability to stimulate ingestion is not restricted to stimulation of the gustatory receptors as was originally assumed when the model was formulated. How that stimulation is mediated and what receptors are involved is unknown. It is a puzzle also because neither of these types of substances would be classified as particularly palatable by humans especially at the high concentrations ingested avidly by the rat. The starches are at most slightly sweet, and a mouthful of corn oil would be rejected quickly by most of us.

4. Evaluation of the Stimulus

To study how the evaluation of the input, the element *p* in *gpd,* influences the overall value of *gpd* it is necessary to test the same group of rats with a constant gustatory stimulus and vary the internal state of the animal. In our first attempt to do this we reported that lowering blood glucose concentrations by peripherally administered insulin increased the value of *gpd* measured by the brief contact method when the rats were tested with five different concentrations of glucose (0.015M to 0.12M) (Davis and Levine, 1977). Because the experiment was a within subject design and because the rats were tested on the same concentrations of glucose under a normo- and hypoglycemic state, we were able to attribute the difference in the value of *gpd* at each concentration to a difference in the value of *p,* the evaluation of the glucose stimulus. The reasoning behind that study was that by making the rats hypoglycemic the evaluation of the taste of glucose would be enhanced, increasing the value of *p* and thus the value of *gpd* as a whole. Hypoglycemia increased the number of licks during the 3-min brief contact tests with each of the five glucose concentrations. However, it increased the number of licks when the rats were tested with distilled water as well, raising the possibility that hypoglycemia caused a generalized excitatory effect on the ingestive mechanism. These results, therefore, are not unambiguous with respect to the hypothesis.

Less ambiguous results have been reported since then. Breslin and his colleagues (1993) recently reported a study supporting the idea that Na depletion, induced by furosemide, increases the evaluation of NaCl. Different groups of rats were tested on different concentrations of NaCl, but each rat was tested on a particular solution in an Na replete and deplete state so any difference the initial lick rate recorded under the two states can be attributed to variations in the parameter *p* and not in the parameters *g* or *d.* The results, shown in Figure 8, were dramatic and convincing. The number of licks during the first 3 min of testing were increased by a factor of at least 3 when the rats were tested in a state of sodium deficiency.

It has also been reported that depriving rats of food increases the sham intake of sucrose, maltose, glucose, and fructose (Joyner *et al.,* 1985). In that study the sham feeding method was used as a measure of *gpd,* and the rats were tested under both nondeprived and food-deprived conditions on the same solutions. Thus the results with maltose, shown in Figure 7, of a much greater intake when food-deprived at every concentration supports the idea that food deprivation increases the value of *p*. Depriving the rats for food for 17 hr increased the value of *gpd* measured in this way by roughly a factor of 4 at each of the 4 concentrations tested.

Each of these three cases show that the value of *p* can be increased by creating a deficit in the internal physiological state of the animal. There is also evidence that the value of *p* can be altered by conditioning. Carmen Perez and I showed this in an unpublished study that examined the changes in intake and initial responsiveness to a flavored test solution that occurred during recovery from a conditioned taste aversion. Such an aversion occurs presumably because of an alteration in the evaluation of the flavor of the test solution because the gustatory stimulation provided by it is the same after as before conditioning. We tested this idea by first adapting rats to a 6mM Na saccharin solution for 10 days, by which time intake was stable. They were then offered that solution flavored with grape Kool Aid on two consecutive 20-min tests, and immediately following those tests they were given an i.p. injection of 0.2M LiCl. A control group was treated in the same way except that they received an isotonic saline injection on the two training tests when the experimental animals were injected with LiCl. Both groups were then offered the grape flavored Kool Aid for nine additional tests without the LiCl or NaCl injections. Figure 10A shows the volumes ingested by the two groups on the 11 test days of the experiment. Figure 10B shows the number of licks that occurred during the first minute of the tests.

The development of the conditioned taste aversion is apparent in the decline in the volume ingested on the two days following the injections with LiCl (tests 2 and 3). From test 4 on the gradual extinction of the aversion, when ingestion was no longer followed by the LiCl injection, is apparent in the daily increasing volume ingested. Because the same rats were being tested with the same test solution, individual differences were held constant, the parameter *d* was constant across tests, and, because the test stimulus was the same over the entire series of tests, the parameter *g* was constant over the tests, changes in the initial rate of licking must have reflected, therefore, an alteration in the evaluation of the taste. That is, the variation in the initial rate of licking that occurred across the tests had to have reflected variation only in the value of the parameter *p,* because *g* and *d* remained constant over them.

Carmen Perez and I have also shown that the evaluation of a test stimulus can be altered by conditioning based on postingestional stimulation (Davis and Perez, 1993a). Earlier Barbara Collins, Michael Levine, and I (Davis *et al.,* 1975) had shown that adding mannitol to a saccharin + glucose solution decreased intake in proportion to the mannitol concentration of the mixture by increasing the magnitude of a negative feedback signal from the gastrointestinal tract, not because of

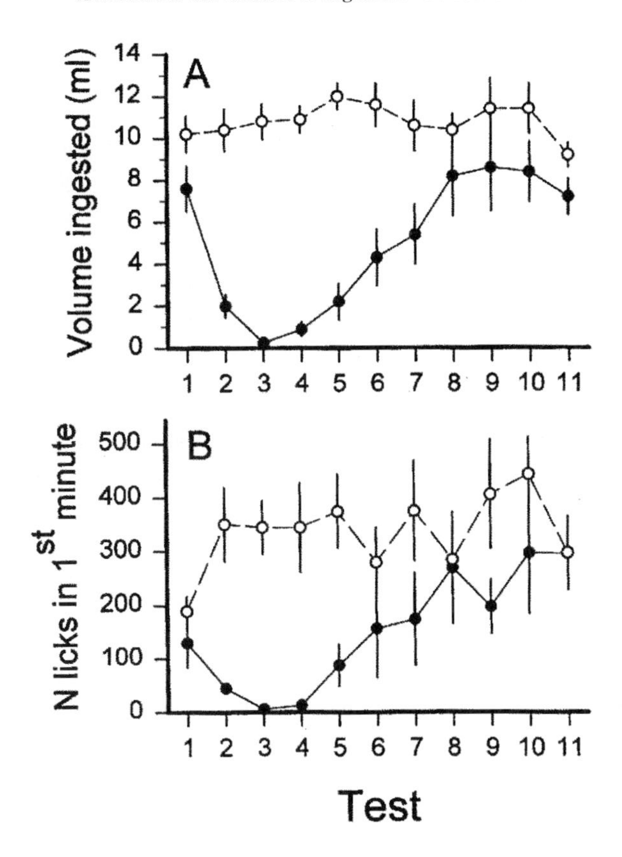

FIGURE 10 Volume ingested (A) and number of licks during the first minute (B) on successive 30-min tests. Filled circles are for rats that received an injection of LiCl immediately following the first two tests. The open circles are for rats that received an injection of isotonic saline immediately following the first two tests.

an alteration in the flavor of the test solution. Because mannitol, a sugar alcohol, is not absorbed from the intestine, adding it to a test solution will lead to an accumulation of fluid in the gastrointestinal tract (Davis and Collins, 1978) creating, we proposed, a negative feedback signal that would reduce meal size.

As discussed earlier in this chapter, we knew that some postingestive consequence of ingestion can serve as an unconditioned stimulus for the conditioning of the rate of licking early in the meal. Because mannitol had clearly defined postingestive negative feedback effects, we examined the possibility that it could serve as an unconditioned stimulus for the conditioning of the rate of licking at the beginning of a meal. To test this idea we trained two groups of rats to ingest a saccharin + glucose solution during 8 daily 30-min test sessions. For one of the groups the test solution during this training period contained 0.3M mannitol; the other solution was mannitol free. During the last four days of this training phase, the intake of the group ingesting the mannitol-containing solution was approximately three time less than the intake of the group ingesting the mannitol free solution (Figure 11A). This was because the initial rate of licking was lower and the rate of decay of the intake curve greater for the group ingesting the mannitol-containing solution than for the group ingesting the mannitol-free solution.

For the next, and remaining three tests, the mannitol was removed from the test solution of one group and added to the test solution of the other. The flavors of the test solutions, however, remained the same. On the first test with these new solutions the rate of licking during the first minute remained the same as it had been on the preceding test (Figure 11B). The removal of mannitol from the test solution of one group and the addition of it to the test solution of the other for the first time had no detectable effect on the initial rate of licking. The initial rate of licking was controlled entirely by the effects of their prior experience with the mannitol-containing and mannitol-free test solutions. The effect of mannitol on the gastrointestinal tract, however, was clearly evident in a very large increase in the rate of decay of the rate of licking by the group ingesting the test solution that now contained mannitol and a very large decrease in the rate of decline in the rate of licking in the group now ingesting the mannitol-free solution for the first time (Figure 11C).

On the second and remaining two tests with these test solutions there was a progressive increase in the initial rate of licking in the group now ingesting the mannitol-free solution and a progressive decrease in the initial rate of licking the test solution that now contained mannitol for the first time (Figure 11B). This demonstrated that the low initial rate of licking could be extinguished in one group and conditioned in the other. The change in the initial rate of licking must have resulted from a change in the evaluation parameter, p, because the conditioning and extinction could be demonstrated in the same animals, that is, d was a constant. We were able to assume that g was a constant because when the mannitol was switched from the solution tasted by one group and introduced into the solution tasted by the other group there was no detectable effect on the initial rate of ingestion. It was only after one association of the flavor with the postingestive effects of mannitol that an effect on the initial rate of licking could be detected.

In a study that followed from the one just described Perez, Kung, and I (Davis *et al.,* 1994a) showed that the same conditioned control of the initial rate of ingestion could be demonstrated by adding Acarbose, which prevents hydrolysis of disaccharides, to a flavored 0.3M sucrose solution. In this study we use a 0.3M sucrose solution as the basic test solution to which 200 mg L^{-1} of Acarbose was added instead of mannitol as in the previous experiment. The design was similar in that different groups of rats were trained with and without Acarbose in the sucrose solution, and after three 30-min tests the Acarbose was removed from the solution ingested by one group and was added to the solution ingested by the other. On the 30-min ingestion tests during the training trials the initial rate of in-

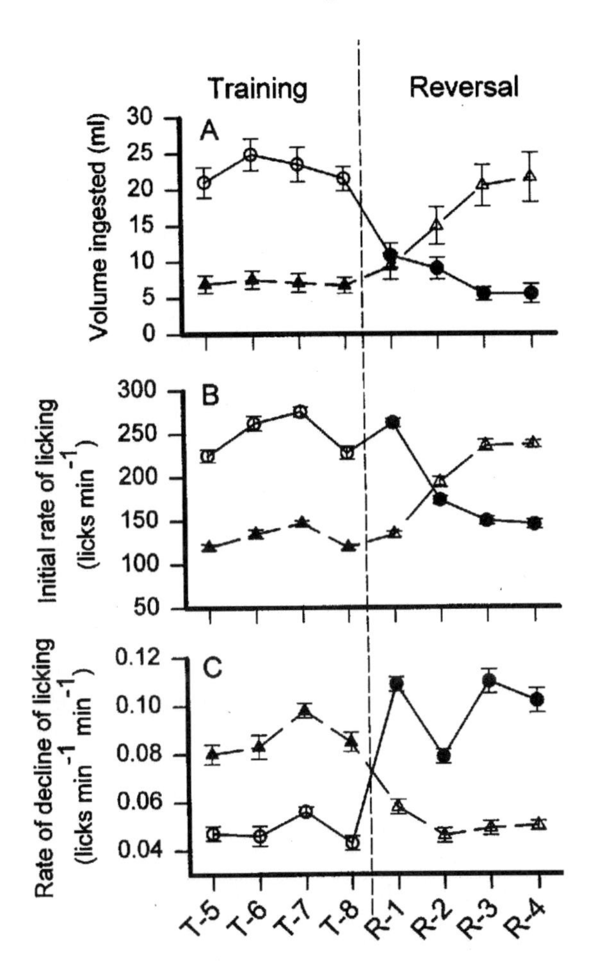

FIGURE 11 Volume ingested (A), initial rate of licking (B), and rate of decline of rate of licking (C) during the last four training trials and the following four reversal tests. Open circles refer to mannitol-free solutions; filled circles refer to mannitol-containing solutions. Redrawn from Davis and Perez (1993a) by permission.

gestion became significantly less for the group ingesting the solution containing Acarbose than for the group ingesting the Acarbose-free solution. When the Acarbose was switched on the first reversal test there was no change in the initial rate of licking. On the following two tests, however, the initial rate of licking of the group now ingesting the Acarbose-containing solution decreased and the initial rate of licking of the group ingesting the Acarbose-free solution increased. The cross-over was complete in three tests. The results of this experiment on the volume ingested, initial rate of licking, and rate of change in the rate of licking were almost identical to those we obtained in the mannitol study shown in Figure 11.

To assure ourselves that the conditioning effect was based on fluid accumulation in the gastrointestinal tract, we measured the volume in the stomach and the small and large intestines in two groups, one having ingested the sucrose solution with Acarbose in it and the other without it. There was no difference between the two groups in the volumes recovered from the stomachs of the rats in the two groups, but there was significantly more fluid recovered from both the small and large intestines of the rats that had ingested the Acarbose-containing sucrose than from the intestines of the rats that had ingested the Acarbose-free sucrose (Davis *et al.,* 1994a).

As in the previous study the same rats were tested on the Acarbose-containing and the Acarbose-free sucrose solutions so the parameter d was constant across conditions. Furthermore, there was no evidence that the rats could taste the Acarbose because the concentration was small and there was no detectable effect on the initial rate of licking when the Acarbose was switched from one test solution to the other, so we could assume that g was constant. The change in the initial rate of licking, gpd, must therefore have been due to a change in the evaluation parameter, p.

In summary, two different types of experiments support the need for the inclusion of the parameter p as a contributor to the value of the initial rate of licking, gpd. Three studies support the idea that the depletion of a mineral or nutrient can increase the initial responsiveness to a test solution. Because the parameters g and d were the same on the two tests and the animals and the test solutions were the same, the difference in the number of licks on the two tests can be modeled as an amplification of the input signal to the ingestion controller.

The other type of study provides evidence for the idea that conditioning can diminish the magnitude of the input signal to the ingestion controller in the CNS. Both of these types of studies support the need for the inclusion of a parameter p in the aggregate input parameter gpd proposed in the model. It should be noted, however, that, although these studies support this aspect of the model, they do not rule out other alternatives. It is possible that these states, or condition-based influences on the ingestive behavior of the rat, operate on the ingestion controller directly or influence the motor output from, rather than afferent input to, the ingestion control system. However, what can be said in favor of the model is that the results do not rule out the inclusion of an evaluation parameter that modifies the input to the ingestion control system of the rat.

C. Shape of the Rate of Licking Functions

One of the problems with the model is that the model predicts that the rate of ingestion during a test will have the form of a continuous exponential decay. It has been known, however, since 1952 with Stellar and Hill's detailed description of the licking behavior of the rat (Stellar and Hill, 1952) that, strictly speaking, the licking behavior of the rat should not be modeled as a continuous process. Rather, licking behavior has the characteristics of an all-or-none activity, with bursts of licking at a high constant rate separated by pauses of varying duration (Stellar and

Hill, 1952). There is an obvious discrepancy here because the behavior of the rat is discrete, and the model describes it as continuous.

The solution I have adopted for this problem is to interpret the ingestion rate functions $l(t)$, and its integral $C(t)$, as describing the trends in the changes of the average rate of ingestion rather than the details of the moment by moment ingestive activity of the animal. This macroscopic approach ignores the details of the individual ingestive acts and focuses instead on the changes in the rate of licking averaged over a finite period of time—usually 1 min. This macroscopic approach ignores the details of the individual ingestive acts and focuses instead on two features of the ingestive behavior of the rat. One is the average rate of ingestion at the beginning of the test, and the second is the change in the average rate of ingestion during the test. According to the model, this gives us information about the effectiveness of the test solution in stimulating ingestive behavior and the magnitude of the postingestive negative feedback signal generated by it, two of the important variables that determine how much will be ingested during the test and the ones with which the model was designed to deal. Other types of information about the mechanisms of control of ingestive behavior can be extracted from a microscopic analysis of the intervals between successive licks (Davis, 1996b; Davis and Smith, 1992).

Given this interpretation of $l(t)$ and $C(t)$, a relevant question is, how well do these functions fit the data? In our *Psychological Review* paper (Davis and Levine, 1977) we addressed this question by fitting the cumulative exponential function, $C(t) = (gp)(rk)^{-1}\,(1\text{-}e^{-drkt})$, to intake and licking data from a variety of different experiments either carried out by ourselves or by others. The test solutions were milk under real and sham feeding conditions, three different concentrations of glucose and sucrose, a glucose solution adulterated with quinine, and a mixture of saccharin and glucose with various concentrations of mannitol added to it. In each case this function fit the data well as assessed by visual inspection. There were no significant deviations of the data from the fitted curves.

In the same year of the publication of our *Psychological Review* paper McCleery (1977) reported a study specifically designed to determine the best fitting curve for intake functions. He selected three functions for comparison: a power function, $N = At^k + C$, the three parameter cumulative exponential function, $N = C - Ae^{-kt}$, similar to the $C(t)$ function predicted by our model, and a rectangular hyperbola, $N = A - C(K + 1)^{-1}$. He showed, by analyzing the residual least-squares errors, that although all could fit the data reasonably well, the exponential function fit the data best. In his study he used rats in a Skinner box bar pressing for 45 mg Noyes pellets, so our model, which was intended to describe the ingestive behavior of rats ingestig liquids, is not strictly applicable. Nevertheless, if it is assumed that the process of satiation by negative feedback is the same in both cases the shape of the intake functions should be similar. Thus, independent of us, a study designed specifically to find the best fitting function for the cumulative intake function of rats concluded, on empirical grounds, that the function generated theoretically by our model was the best fit to the data.

More recently in a study designed to investigate the effect of food deprivation on the ingestive behavior of rats ingesting five different concentrations of sucrose ranging from 0.05M to 0.8M (Davis and Perez, 1993b), the rate function, $l(t) = gpe^{-rkt}$, was fit to a curve generated by calculating the average rate of licking (licks min^{-1}) averaged over successive 2 min intervals of the 30-min test session. Under both 0 hr and 17 hr of food deprivation the fits to the data were good, as indicated by the fact that the standard errors of estimate for the parameters were small.

In a study designed to investigate the shape of the intake curves of rats ingesting a wide variety of concentrations of corn oil emulsions (Davis *et al.*, 1995), the Weibull function, a generalization of the exponential, was used successfully to fit the rate of licking functions. The Weibull function, $y = ae^{(-bt)^c}$, rather than the simple exponential was chosen because, at the expense of estimating just one more parameter, c, it permitted a quantitative evaluation of how much the function deviated from a true exponential, the function demanded by the model. Figure 12 shows these fits for a low and high concentration of corn oil.

The fits of this function to the rate of licking curves for corn oil were good as assessed by visual inspection and by the fact that the corrected r^2 were in no case less than 0.91. (The correlation coefficients were not reported in that paper but were obtained in a subsequent analysis of those data.) The estimates of the shape parameter, c, ranged between 1.5 and 0.5, with one unexplained outlier at a corn oil concentration of 0.25%. The estimates tended to be somewhat greater than 1

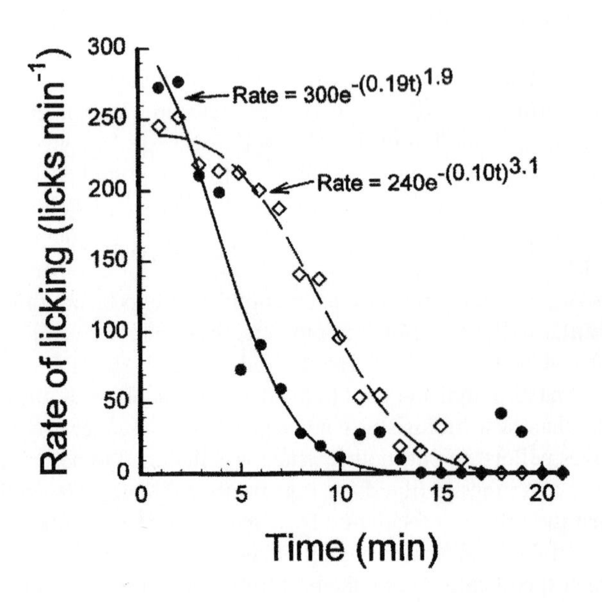

FIGURE 12 Rate of licking on the last real feeding test (filled circles) before seven cycles of two real feeding tests followed by a sham feeding test and the last of the real feeding tests (open diamonds).

for corn oil concentrations below about 8% and close to or slightly below 1 at the higher concentrations. In other words, at the lower concentrations of corn oil the rate of licking did not decline as rapidly early in the test as predicted by the model, and at the higher concentrations it declined more rapidly than predicted. This is a discrepancy that indicates a problem with the model, a problem that can be resolved by introducing a threshold for the activation of negative feedback.

The model predicts that the magnitude of the negative feedback signal from gastrointestinal accumulation is reflected in the rate of decay of licking during the test. If the rate of licking does not begin to decay until sometime after the beginning of a test, the implication is that the occurrence of negative feedback does not appear until then. In other words, there is a delay between when the rat begins to ingest the test solution and the appearance of postingestional negative feedback. This would be reflected in the shape parameter of the rate of licking function being greater than 1. A shape parameter less than 1, however, would suggest the accelerated onset of a negative feedback signal. In other words, an increasing value of the shape parameter, c, with decreasing corn oil concentrations is consistent with the idea of a threshold for the activation of postingestional negative feedback, which depends on the concentration of corn oil.

A threshold for the activation of negative feedback was considered when we designed the original model but was not included in it for reasons of simplicity; it would have introduced a nonlinearity that would have complicated the model beyond our initial intent. Furthermore, because of this nonlinearity, it would not have been possible to derive an equation to describe the changes in the rate of licking during a test. This would have made it impossible to test the model quantitatively. Nevertheless, it is still possible to explore the need for a threshold for negative feedback by examining the rate of licking functions in situations where a delay in the initiation of negative feedback might be expected, and in situations where the opposite might be expected. The Weibull function will be used to do this, but it should be clear that this is entirely an empirical exercise, a post hoc patch, not a theoretically derived prediction from the model. It was the pattern of results that arose from this approach that suggested the use of it in detecting thresholds for the activation of negative feedback.

Whatever the nature of negative feedback from postingestional stimulation, neural, or humoral, it is concentration dependent, as will be discussed. It, therefore, seems reasonable to assume that more fluid of a low concentration needs to be ingested before enough has accumulated in the gastrointestinal tract for it to activate any form of negative feedback, chemical or neural. An unpublished study designed to study the effects of the dilution of sweetened condensed milk on the change in the rate of licking during the meal provided clear evidence in support of this idea. Rats were tested for 30 min after 4 hr of food deprivation on 1:1, 2:1, 4:1, 8:1, and 16:1 dilutions of Borden's sweetened condensed milk (milk:distilled water). They were tested on each dilution until the volume ingested during the test was stable and then were tested on the next more diluted solution. The functions describing the rate of licking calculated at 1-min intervals obtained from the last test with each dilution were fit by the least squares method to the Weibull function. The estimates of the three parameters of the Weibull function are given in Table 1. There it can be seen that with dilutions of 1:1 and 2:1 the estimates of the shape parameter, *c,* were roughly the same, and at 4:1 and lower they were much larger. This trend in the estimates of this parameter is consistent with the idea that when ingesting less concentrated solutions containing fat it takes longer for a negative feedback signal to reach threshold.

A different experiment has produced results quite consistent with this idea. In that study postgastric stimulation, but not gastric stimulation, was eliminated by confining all of the fluid ingested during the test to the stomach by inflating a cuff around the pylorus (Davis *et al.,* 1993). The rats were given seven cycles of two real feeding tests followed by a sham feeding test. On all of the tests the pyloric cuff was inflated, and on the real feeding tests the contents of the stomach were aspirated before the cuff was released. As a result, on the sham feeding tests there was no gastric accumulation and no postgastric stimulation, and on the real feeding tests there was maximal gastric stimulation, but no postgastric stimulation.

The results were that intake increased significantly by almost 60% from the first to the last real feeding test (Figure 5). The reason for the increase was that the rate of licking did not begin to decline until later and later in the test. This could be seen in the estimate of the shape parameter of the Weibull function, which increased by 60% from 1.9 on the first real feeding test to 3.1 on the last. Figure 12 shows the rate of licking curves for these feeding tests along with the Weibull functions fit to the data.

Our interpretation of those data was that conditioned control of the rate of licking early in the test was extinguished because of the absence of postgastric stimulation permitting the rate of licking during the early part of the test to increase. However, because on these real feeding tests all of the ingested fluid was confined to the stomach, gastric distention eventually provided an unconditioned physiological negative feedback signal that terminated the meal. Negative feedback was delayed as the rats gained experience with this procedure because the conditioned component of negative feedback was extinguished (Davis *et al.,* 1993) leaving only a gastric component, which was activated toward the end of the meal when accumulating fluid began to approach the capacity of the stomach.

Supporting this view are the results from a recently reported study (Davis *et al.,* 1997) where we showed that when ingested fluid is confined to the stomach by closing a pyloric cuff during a meal there was no evidence of an effect of gastric filling on the rate of licking until about 6 or 7 min of licking has occurred resulting in the accumulation of about 7 ml to 8 ml in the stomach. That is, it takes 6 to 7 min for negative feedback to appear in the kind of test situation where postgastric stimulation cannot occur and negative feedback is derived entirely from the stomach. Note in Figure 12 that the curve describing the change in the rate of licking on the last real feeding test (open diamonds) did not begin to decrease significantly until about 5 min after the test began. It seems likely that it was at this point that enough fluid had accumulated in the stomach to generate a negative feedback signal sufficient in magnitude to slow the rate of licking.

The opposite case, where the value of the shape parameter of the Weibull function was decreased by an experimental manipulation, occurred in a study where we reported the effects of total subdiaphragmatic vagotomy on the ingestion of milk (Davis *et al.,* 1994b). Vaogotomy reduced the intake of milk by about 85% and the duration of the meal by about 40%. The major effect of vagotomy on intake occurred, however, because the rate of decline of the rate of licking was much more rapid than in intact rats. The half-life of the rate of licking functions in the vagotomized rats decreased from a preoperative value of about 11 min to a postoperative value of about 4 min. This occurred because of an increase in the value of the rate of decay of licking parameter from about 0.07 licks min^{-1} to about 0.21 licks min^{-1} and decrease in the value of the shape parameter of the Weibull function from a baseline value of about 1.8 to about 1.2 postvagotomy. We attributed that result to a premature and intensified negative feedback signal from accelerated gastric clearance of liquids that is known to follow total subdiaphragmatic vagotomy (Alvarez, 1948; Anita *et al.,* 1951; Schwartz *et al.,* 1993).

In all the experiments previously discussed the Weibull function fit the data well and, more to the point, the shape parameter varied in away consistent with the idea that it reflects the time of onset of negative feedback. In two studies the value of the shape parameter decresed as the concentration of the test solutions containing oil increased. In another study, where the negative feedback signal is likely to have been reduced by eliminating postgastric stimulation, the value of the shape parameter increased dramatically. And, finally, in that case where abnormally rapid gastric emptying, a syndrome that follows vagotomy is likely to have occurred, the value of the shape parameter decreased substantially.

In the experiments mentioned previously the rate function *l(t),* the Weibull function, or the cumulative form, *C(t),* of the predicted function were fit to data obtained by averaging across a group of rats. This raises the possibility that the form of the curve was created by the process of averaging. To evaluate this possibility, curve fitting was done in two ways in two recently reported studies (Davis, 1996a; Breslin *et al.,* 1996). In one study I conducted with Paul Breslin and Rachel Rosenak (Breslin *et al.,* 1996) investigating the effects of varying the concentration of saccharin in a saccharin + glucose mixture, we fit the cumulative, *C(t),* function to both the individual animal intake data and to curves obtained by averaging the individual curves of the 12 animals used in the experiment. The estimates of the two parameters obtained from the averaged curves and the averages of the estimates obtained from the individual animal curves agreed closely. This is supported by the fact that in all but one case the standard errors of estimates obtained from the fits to the grouped data and the standard errors of the means obtained by averaging the individual estimates overlapped. In the odd case they barely missed overlapping.

In the other study (Davis, 1996a) the same analysis was applied to data from a study that investigated the effect of variation in Polycose concentration on the rat's ingestive behavior. The results were the same, the averages of the estimates of the parameters obtained from the individual rate of licking curves were not significantly different from those obtained from fitting the averaged rate of licking curves. These two studies indicate that, at least for these data and types of test solutions, the negative exponential shape of the average rate of licking function is not an artifact of averaging the individual curves.

Although all of these results support the prediction of the model that the rate of licking curve is exponential in shape, results are accumulating that this may not be true for all testing situations. In particular, a number of laboratories have reported that within individual meals recorded over 24-hr periods in undisturbed rats ingesting a single diet the rate of ingestion is constant; there is no sign of a decline in the rate of ingestion during the meal (Rushing *et al.,* 1997; Garry Schwartz, Tim Moran, and George Collier, personal communications). Why there is such a big difference in the shape of the ingestion curves in these two different types of testing situations is not at all clear because the reasons for the differences have not yet been studied in any detail. There are certainly many differences in the types of variables that are likely to be playing a role in controlling intake in the two situations, but until they have been given some experimental attention further speculation is pointless.

IV. Summary

In summary it is clear that the model was, with a few exceptions, wrong in its specific predictions. The predictions that intake would be maximum and depend only on *gpd* the first time a rat was tested on sham feeding, and that the rate of licking within a sham feeding test would be constant were wrong. Although the idea that negative feedback depended on intestinal fill has been supported by more recent research, the idea of basing negative feedback exclusively on intestinal fill or on the intestine alone was clearly wrong. Although the prediction that the initial rate of licking is a function of the magnitude of gustatory stimulation has been well supported, it is now clear that other types of orophyrangeal stimulation can affect the initial rate of licking as well. The need for a parameter p to account for variations in the initial rate of licking dependent on variations in the state of the animal has been supported, to the extent that it has been studied. And, finally, although the shape of the rate of licking curve can be fit in many cases by a simple exponential function, the Weibull function, usually does a better job, and it is becoming increasingly clear that in other types of test situations there is no decay at all in the rate of ingestion during the meal.

This to me is not the least bit discouraging. We never expected the model to be correct in any really precise sense. It was designed to organize our thoughts about the controls of meal size and to generate predictions that could be tested in a quantifiably specific way. This has been done to some extent and the results of those tests have helped some of us pinpoint where our ideas and assumptions about the variables that control the intake of liquids by rats are wrong. The negative results have forced us to ask some questions that we might not have asked if the discrepancy between prediction and outcome were not so clearly apparent.

The value of the model, as I see it, has been in the quantitative specificity of the predictions it made. Many were wrong, but discovering that has served the purpose of forcing revisions in the model. These have been incorporated in the diagram shown in Figure 13. A major revision is the inclusion of conditioning, which I have done by introducing the box labeled "Memory." New pathways, shown as dashed lines, leading to this box represent input from conditioned and unconditioned stimuli. In some cases there is clear evidence to support the pathway, oral stimulation, for example, as a conditioned stimulus for taste aversion learning. In other cases, gastric stimulation, for example, as the source of unconditioned stimulation for conditioned oral control of ingestion, the pathway, while possible, is speculative. The box labeled "Threshold" is there because of its theoretical necessity and the fact that the shape parameter of the Weibull function fit to the rate of licking curves are often greater than 1 suggesting a delay in the onset of negative feedback.

This revised model is an improvement over the original one, but like the original model, it only incorporates what Smith (1996) has called the direct controls of

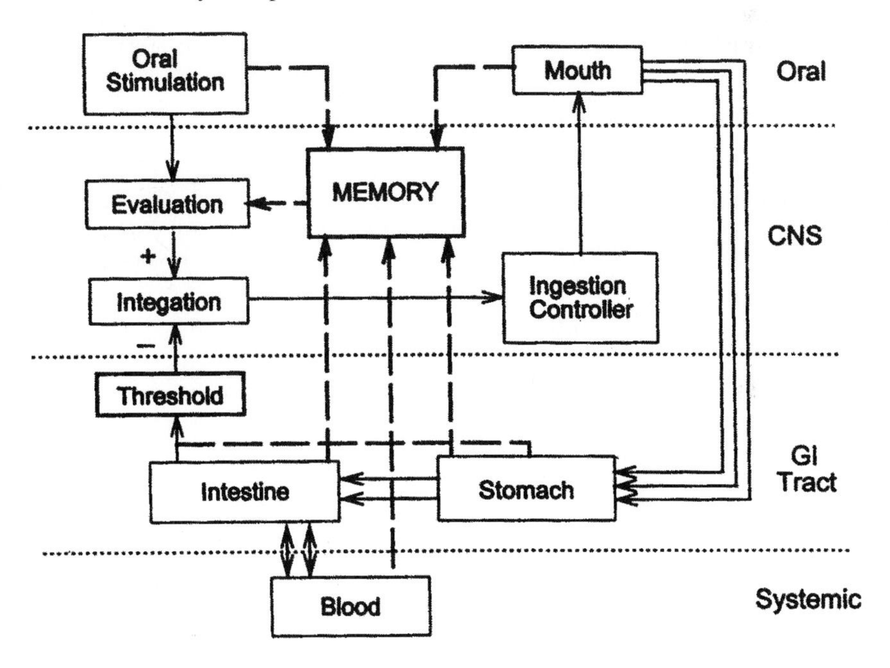

FIGURE 13 Revised schematic of the Davis and Levine model of the control of food intake. Dashed lines show new pathways suggested by research conducted since the publication of the original model.

ingestion. Much that we now know about feeding behavior is still not represented in it largely because we do not know where in the periphery or CNS many variables that influence intake exert their influence. How, for example, does food deprivation influence intake? What receptors respond to energy deficit, and where in the ingestive control system do they exert their influence? Some neuropeptides such as NPY are well known to stimulate appetite in the rat, but how they do it and where they act on the ingestive control system remains a mystery. When these so called indirect controls of meal size (Smith, 1996) are better understood, they may find a place in a more elaborate quantitatively explicit model such as the one described here. Until then they will be difficult to quantify.

Acknowledgments

I am indebted first of all to Dr. Gerard P. Smith who has been encouraging me to write this review for some time. Without his persistence, I never would have started it. I am also indebted to him for a long and very fruitful collaboration, begun more than 10 years ago and continuing today. It will be clear to any reader of this review that many of the significant revisions in the model, and the more interesting discoveries, especially with respect to sham feeding, are the result of a close collaboration between us.

I am also indebted to my many other collaborators over the years who have made many contributions of one kind or another to the development of this model. Of these I especially want to thank, Barbara J. Collins and Mike Levine, who were my original collaborators in the development of the model. Their interests have taken them on different paths since then, but I remember fondly the fun we had together in a seminar that Mike conducted on control theory, which gave birth to this model. David Wirtshafter, another member of that seminar, who, at tmes had more faith in the model than I did, has contributed in many was to my thinking during the years of informal dicussions. We never did any experiments together specificially related to this model, but many discussions with him have clarified my thinking on many problems related to it.

The others I want to acknowledge are those who have been of great help to me in conducting the experiments. They are Dora Kung, Judy Miesner, Carmen Perez, Rachel Rosenak, and Jennifer Sayler. Their mastery of the mechanics of the experiments, their surgical techniques, careful attention to detail, and complete reliability made it possible for me to be able to write much of this chapter, for without their help there would have been much less to write about.

References

Alvarez, W. C. (1948). Sixty years of vagotomy: A review article of some 200 articles. *Gastroenterology* **10,** 413–441.

Anita, F., Rosiere, C. E., Robertson, C., & Grossman, M. I. (1951). Effects of vagotomy on gastric secretion and emptying time in dogs. *American Journal of Physiology* **166,** 470–479.

Bartoshuk, L. M. (1968). Water taste in man. *Perception Psychophysics,* **3:** 69–72.

Bartoshuk, L. M. (1987). Is sweetness unitary? In J. Dobbing (Ed.), *Sweetness* (pp. 33–48). London: Springer.

Bernstein, I. L., & Vitiello, M. V. (1978). The small intestine and the control of meal patterns in the rat. *Physiology and Behavior* **20,** 417–421.

Booth, D. A. (1972). Conditioned Satiety in the rat. *Journal of Comparative Physiological Psychology* **81,** 457–471.
Booth, D. A., & Davis, J. D. (1973). Gastrointestinal factors in the acquisition of oral sensory control of satiation. *Physiology and Behavior* **11,** 23–29.
Breslin, P. A., Davis, J. D., & Rosenak, R. (1996). Saccharin increases the effectiveness of glucose in stimulating ingestion in rats but has little effect on negative feedback. *Physiology and Behavior* **60,** 411–416.
Breslin, P. A. S., Kaplan, J. M., Spector, A. C., Zambito, C. M., & Grill, H. J. (1993). Lick rate analysis of sodium taste-state combinations. *American Journal of Physiology* **264,** R312–R318.
Campbell, C. S., & Davis, J. D. (1974). Licking rate of rats is reduced by intraduodenal and intraportal glucose infusion. *Physiology and Behavior* **12,** 357–365.
Collins, B. J., & Davis, J. D. (1978). Long term inhibition of intake by mannitol. *Physiology and Behavior* **21,** 957–965.
Davis, J. D. (1973). The effectiveness of some sugars in stimulating licking behavior in the rat. *Physiology and Behavior* **11,** 39–45.
Davis, J. D. (1996a). Microstructural analysis of the ingestive behavior of the rat ingesting Polycose. *Physiology and Behavior* **60,** 1557–1563.
Davis, J. D. (1996b). Deterministic and probabilistic control of the behavior of rats ingesting liquid diets. *American Journal of Physiology* **270,** R793–R800.
Davis, J. D., & Campbell, C. S. (1973). Peripheral control of meal size in the rat: Effect of sham feeding on meal size and drinking rate. *Journal of Comparative Physiological Psychology* **83,** 379–387.
Davis, J. D., & Collins, B. J. (1978). Distention of the small intestine, satiety, and the control of food intake. *American Journal of Clinical Nutrition* **31,** S255–S258.
Davis, J. D., Collins, B. J., & Levine, M. W. (1975). Peripheral control of drinking: Gastrointestinal filling as a negative feedback signal: A theoretical and experimental analysis. *Journal of Comparative Physiological Psychology* **89,** 985–1003.
Davis, J. D., Kung, T. M., & Rosenak, R. (1995). Interaction between orosensory and postingestional stimulation in the control of corn oil intake by rats. *Physiology and Behavior* **57**(6), 1081–1087.
Davis, J. D., & Levine, M. (1977). A model for the control of ingestion. *Psychology Review* **89,** 379–412.
Davis, J. D., & Perez, M. C. (1993a). The acquired control of ingestive behavior in the rat by flavor-associated postingestional stimulation. *Physiology and Behavior* **54,** 1221–1226.
Davis, J. D., & Perez, M. C. (1993b). Food deprivation and palatability-induced microstructural changes in ingestive behavior. *American Journal of Physiology* **264,** R97–R103.
Davis, J. D., Perez, M. C., & Kung, T. M. (1994a). Conditioned control of ingestive behavior of Acarbose induced inhibition of sucrose digestion. *Physiology and Behavior* **55,** 511–518.
Davis, J. D., & Sayler, J. L. (1997). Effects of confining ingested fluid to the stomach on water and saline intake in the rat. *Physiology and Behavior* **61,** 127–130.
Davis, J. D., & Smith, G. P. (1990). Learning to sham feed: Behavioral adjustments to loss of physiological postingestional stimuli. *American Journal of Physiology* **259,** R1228–R1235.
Davis, J. D., & Smith, G. P. (1992). Analysis of the microstructure of the rhythmic tongue movements of rats ingesting maltose and sucrose solutions. *Behavioral Neuroscience* **106,** 217–228.
Davis, J. D., Smith, G. P., & Kung, T. M. (1994b). Abdominal vagotomy alters the structure of the ingestive behavior of rats ingesting liquid diets. *Behavioral Neuroscience* **108,** 767–779.
Davis, J. D., Smith, G. P., & Miesner, J. (1993). Postpyloric stimuli are necessary for the normal control of meal size in real feeding and sham feeding rats. *American Journal of Physiology* **265,** R888–R895.
Davis, J. D., Smith, G. P., & Sayler, J. L. (1997). Reduction of intake in the rat due to gastric filling. *American Journal of Physiology* **272,** R1599–R1606.
Dethier, V. G. (1976). *The hungry fly,* Harvard: Cambridge.
Deutsch, J. A. (1990). Food intake: Gastric factors. In E. M. Stricker (Ed.), *Handbook of behavioral neurobiology,* Vol. 10, (pp. 151–182). New York: Plenum Press.
Ernits, T., & Corbit, J. D. (1973). Taste as a dipsogenic stimulus. *Journal of Comparative Physiological Psychology* **83,** 27–31.

Feigin, M. B., Sclafani, A., & Sundal, S. R. (1987). Species differences in polysaccharide and sugar taste preferences. *Neuroscience and Biobehavioral Review* **11,** 231–240.

Geary, N. (1996). Glucagon and the control of appetite. In P. J. Lefebvre (Ed.), *Handbook of experimental pharmacology,* Vol. 123, Glucagon III (pp. 223–238). Berlin: Springer-Verlag.

Gibbs, J., & Smith, G. P. (1992). Peripheral signals for satiety in animals and humans. In G. H. Anderson & S. H. Kennedy (Eds.), *The biology of feast and famine* (pp. 61–72). New York: Academic Press.

Greenberg, D., Gibbs, J., & Smith, G. P. (1986). Intraduodenal infusions of fat inhibit sham feeding in Zucker rats. *Brain Research Bulletin* **20,** 779–784.

Greenberg, D., Smith, G. P., & Gibbs, J. (1988). Oleic acid inhibits sham feeding when duodenally infused while triolein does not. *Society for Neuroscience Abstracts* **14,** 1196.

Greenberg, D., Smith, G. P., & Gibbs, J. (1990). Intraduodenal infusions of fats elicit satiety in the sham feeding rat. *American Journal of Physiology* **259,** R110–R118.

Hall, W. G. (1973). A remote stomach clamp to evaluate oral and gastric controls of drinking in the rat. *Physiology and Behavior* **11,** 897–901.

Hettinger, T. P., Frank, M. E., & Myers, W. E. (1996). Are the tastes of polycose and monosodium glutamate unique? *Chemical Senses* **21,** 341–347.

Jacobs, H. L. (1961). The osmotic postingestion factor in the regulation of glucose appetite. In M. R. Kare & B. P. Halpern (Eds.), *The physiological and behavioral aspects of taste.* Chicago: University of Chicago Press.

Janowitz, H. D., Hollander, F., Orringer, D., Levy, M. H., Winkelstein, A., Kaufman, M. R., & Margolin, S. G. (1950). A quantitative study of the gastric secretory response to sham feeding in a human subject. *Gastroenterology* **16,** 104–116.

Joyner, K. S., & Smith, G. P. (1986). Differential effects of four sugars on sham feeding by obese and lean Zucker rats. *Eastern Psychological Association (Abstracts)* **57,** 65.

Joyner, K., Smith, G. P., Shindledecker, R., & Pfaffmann, C. (1985). Stimulation of sham feeding in the rat by sucrose, maltose, glucose, and fructose. *Society for Neuroscience Abstracts* **11,** 1223.

Louis-Sylvestre, J. (1978). Relationship between two stages of prandial insulin release in rats. *Journal of Physiology* **4,** E103–E111.

McCleary, R. A. (1953). Taste and post-ingestive factors in specific-hunger behavior. *Journal of Comparative Physiological Psychology* **46,** 411–421.

McCleery, R. H. (1977). On satiation curves. *Animal Behavior* **25,** 1005–1015.

McFarland, D. J. (1971). Feedback mechanisms in animal behavior. London: Academic Press.

Mook, D. G. (1963). Oral and postingestional determinants of the intake of various solutions in rats with esophageal fistulas. *Journal of Comparative and Physiological Psychology* **56,** 645–659.

Mook, D. G., Culberson, R., Gelbart, R. J., & McDonald, K. (1983). Oropharyngeal control of ingestion in rats: Acquisition of sham-drinking patterns. *Behavioral Neuroscience* **97,** 574–584.

Nissenbaum, J. W., & Sclafani, A. (1987). Sham-feeding response of rats to polycose and sucrose. *Neuroscience and Biobehavioral Review* **11,** 215–222.

Pavlov, I. P. (1928). *Lectures on conditioned reflexes* (W. H. Gantt, Trans.) New York: Liveright Publishing Corp.

Phillips, R. J., & Powley, T. L. (1996). Gastric volume rather than nutrient content inhibits food intake. *American Journal of Physiology* **271,** R766–R779.

Powley, T. (1977). The ventromedial hypothalamic syndrome, satiety and a cephalic phase hypothesis. *Psychological Review* **84,** 89–126.

Rauhofer, E. A., Smith, G. P., & Gibbs, J. (1991). Prevention of gastric emptying does not alter food intake. *Proceedings of the Eastern Psychological Association (abstracts)* **62,** 10.

Reidelberger, R. D., Kalogeris, T. J., Leung, P. M. B., & Mendel, V. E. (1983). Postgastric satiety in the sham feeding rat. *American Journal of Physiology* **244,** R872–R881.

Richter, C. P., & Campbell, K. H. (1940). Taste thresholds and taste preferences of rats for five common sugars. *Journal of Nutrition,* **20:** 31–46.

Riggs, D. S. (1970). Control theory and physiological feedback mechanisms. Baltimore, MD: Williams & Wilkins.

Rushing, P. A., Houpt, T. A., Henderson, R., & Gibbs, J. (June, 1997). Microstructural analysis of spontaneous meals in rats maintained on a liquid diet. Paper presented at International Behavioral Neuroscience Society. San Diego, CA. .

Schwartz, G. J., Berkow, G., McHugh, P. R., & Moran, T. H. (1993). Gastric branch vagotomy blocks nutrient and cholecystokinin-induced suppression of gastric emptying. *American Journal of Physiology* **264,** R630–R637.

Sclafani, A. (1987). Carbohydrate taste, appetite, and obesity: An overview. *Neuroscience and Biobehavioral Reviews* **11,** 131–153.

Sclafani, A., & Clyne, A. E. (1987). Hedonic response of rats to polysaccharide and sugar solutions. *Neuroscience and Biobehavioral Reviews* **11,** 173–180.

Sclafani, A., & Nissenbaum, J. W. (1985). Is gastric sham-feeding really sham-feeding? *American Journal of Physiology* **248,** R387–R390.

Shuford, E. H. (1959). Palatability and osmotic pressure of glucose and sucrose solutions as determinants of intake. *Journal of Comparative Physiological Psychology* **51,** 150–153.

Smith, G. P. (1996). The direct and indirect controls of meal size. *Neuroscience and Biobehavioral Reviews* **20,** 41–46.

Smith, J. C., Davis, J. D., & O'Keefe, G. B. (1992). Lack of an order effect in brief contact taste tests with closely spaced test trials. *Physiology and Behavior* **52,** 1107–1111.

Smith, G. P., & Gibbs, J. (1992). The development and proof of the CCK hypothesis of satiety. In C. T. Dourish, S. J. Cooper, S. D. Iversen, & L. L. Iversen (Eds.), Multiple cholecystokinin receptors in the CNS (pp. 166–182). Oxford: Oxford University Press.

Stellar, E., & Hill, J. M. (1952). The rat's rate of drinking as a function of water deprivation. *Journal of Comparative Physiological Psychology* **45,** 96–102.

Swithers-Mulvey, S. E., & Hall, W. G. (1992). Control of ingestion by oral habituation in rat pups. *Behavioral Neuroscience* **106,** 710–717.

Thein, M. P., & Schofield, B. (1959). Release of gastrin from the pyloric antrum following vagal stimulation by sham feeding in dogs. *Journal of Physiology* **148,** 291–305.

Toates, F. (1975). Control theory in biology and experimental psychology. London: Hutchinson Educational.

Van Vort, W., & Smith, G. P. (1983). The relationships between the positive reinforcing and satiating effects of a meal in the rat. *Physiology and Behavior* **30,** 279–284.

Van Vort, W., & Smith, G. P. (1987). Sham feeding experience produces a conditioned increase of meal size. *Appetite* **9,** 21–29.

Walls, E. K., Phillips, R. J., Wang, F. B., Holst, M.-C., & Powley, T. L. (1995). Suppression of meal size by intestinal nutrients is eliminated by celiac vagal deafferentation. *American Journal of Physiology* **269,** R1410–R1419.

Weingarten, H. P., & Kulikovsky, O. T. (1989). Taste-to-postingestive consequence conditioning: Is the rise in sham feeding with repeated experience a learning phenomenon? *Physiology and Behavior* **45,** 471–476.

Weingarten, H. P., & Watson, S. D. (1982). Sham feeding as a procedure for assessing the influence of diet palatability on food intake. *Physiology and Behavior* **28,** 401–407.

Young, P. T. (1966). Hedonic organization and regulation of behavior. *Psychological Review* **73,** 59–86.

Young, P. T. (1967). Palatability: The hedonic response to foodstuffs. In C. F. Code (Ed.), Handbook of physiology (Section 6): Alimentary canal, Vol. 1: Control of food and water intake. Washington, DC: American Physiological Society.

Young, R. C., Gibbs, J., Antin, J., Holt, J., & Smith, G. P. (1974). Absence of satiety during sham feeding in the rat. *Journal of Comparative Physiological Psychology* **87,** 795–800.

Young, P. T., & Trafton, C. L. (1964). Activity contour maps as related to preference in four gustatory stimulus areas of the rat. *Journal of Comparative Physiological Psychology* **58,** 68–75.

Author Index

C

D

R

T

X

Y

Z

Subject Index

ISBN 0-12-542117-6